Springer Undergraduate Mathematics Series

For other titles published in this series, go to
www.springer.com/series/3423

Alan Camina · Barry Lewis

An Introduction to Enumeration

 Springer

Alan Camina
School of Mathematics
University of East Anglia
Norwich NR4 7TJ
UK
a.camina@uea.ac.uk

Barry Lewis
The Mathematical Association
Leicester LE2 3BE
UK
barry@mathscounts.org

ISSN 1615-2085
ISBN 978-0-85729-599-6   e-ISBN 978-0-85729-600-9
DOI 10.1007/978-0-85729-600-9
Springer London Dordrecht Heidelberg New York

British Library Cataloguing in Publication Data
A catalogue record for this book is available from the British Library

Library of Congress Control Number: 2011929345

Mathematics Subject Classification (2010): 05AXX

Cover design: SPI Publisher Services

Printed on acid-free paper

Springer is part of Springer Science+Business Media (www.springer.com)

Preface

This book is aimed at undergraduates interested in discrete mathematics, enumeration or combinatorics. We have chosen the title *An Introduction to Enumeration* because of a single theme which will run through most of the book. The theme is counting using series: sometimes infinite and sometimes finite.

In the beginning, as children, our first introduction to mathematics is counting. It comes as a surprise to realize that this concept involves functions. Offer four children three chocolate bars and the chances are that one will scream, a one–one function being instinctively understood; enumeration understood. But the bigger surprise is that from counting marbles or chocolate biscuits, which is the discrete world, we can, in one simple step, go to the continuous world, and this new world provides powerful insights into the old world. A further surprise is that group theory can be used to count, and it comes into its own when symmetries are present in the configurations we are interested in enumerating. Because of the use of analysis and group theory we would expect this book to be of interest to anyone who has completed a first-year undergraduate course in mathematics. For background to these topics there are the following books in this series: J. M. Howie [8, 9] and G. Smith & O. Tabachnikova [14]. We should also mention the book by Ian Anderson in this series, [1], which overlaps with our book.

It is the use of generating functions to represent an enumerative sequence which is the link to these other branches of mathematics. A sequence arises naturally when we try to answer a question such as, "Given a particular configuration of some objects, how many such objects are there of size r?". We usually suppose that there are u_r of them: the ordered list $\{u_r : r = 1, 2, 3, \cdots\}$ is the sequence of interest. Now attach each u_r to the rth power of z, which is z^r. At this stage we leave the question of what z is a little vague. This leads naturally to the series $\sum_{r=0}^{\infty} u_r z^r$. This object, called a power series, is packed full of enumerative information about the sequence $\{u_r\}$. We access this information through a function that the power series defines. The function, when expanded as a (convergent) Taylor series, has as its coefficients the terms of the sequence $\{u_r\}$. We must be careful about the distinction between the series and the function. There is, however, a formal way of thinking about power series which avoids these issues. There are interesting discussions of these ideas, though from an abstract viewpoint, in D. Zelberger's article in [5], H. S. Wilf [18, 17] and I. Nevin [12]. We do not concern ourselves with such introspection (important though it is): the time for that must follow this book.

One of the ways in which this book differs from many others is that, wherever possible, it uses diagrams that portray subtle and slippery strands of an argument graphically so that they may be immediately grasped.

The first chapter sets the scene by asking in how many ways k objects can be chosen from r objects. The answer is not straightforward and depends on a number of constraints. Is the order in which the objects are chosen significant? In some gambling games this can be crucial. Are the objects distinguishable? Can they be selected more than once – so that there is an inexhaustible supply of them? Such questions perfectly describe the objective of the book: given an object of a certain size and configuration, how many of them are there? What are the relations between objects of different sizes, and how are they related to other objects? We introduce some simple techniques and then explore such objects, chosen from many areas of mathematics – geometry, sets, matrices, functions, groups, symmetries, permutations, paths and partitions. Time and time again, we use the same tools and the same techniques to unearth powerful results.

Chapter 4 introduces the next major theme of the book: group theory and its use in enumeration. Group theory began life as the study of symmetry. So as soon as the counting needs to take into account things which might appear to be the same, group theory naturally arises.

The other chapters take a particular problem and use this to motivate, guide and direct the development of the subject:

 (i) the number of ways of giving change in a particular coinage for the purchase of an item when a given amount is proffered;

 (ii) the number of different ways that a cube can be coloured with three colours;

(iii) the number of different paths there are in a grid made up from integer coordinate points;

(iv) the number of ways that a particular score may be gained when multiple dice are thrown;

 (v) the number of ways that subsets of a set can be chosen if the chosen subsets must not contain consecutive elements.

In each case we explore the enumeration of these configurations using tools and techniques that are progressively developed.

Enumeration leads to interesting problems and in this book we develop the theory, linked always to the ideas, tools and techniques on which it relies. We include many exercises (all with full answers) designed to motivate, consolidate and extend mathematical engagement so that the ideas are seamlessly absorbed and mastered.

A fascinating resource for sequences is the web site developed by N. J. A. Sloane [13]. We include in the bibliography some books not mentioned in the text which the reader might find interesting.

Alan Camina and Barry Lewis, January 2011

Contents

1

What Is Enumeration?

Enumerative mathematics (also commonly called combinatorics) is concerned with arrangements of a given set of objects according to precise rules. The aim is to count the number of possible arrangements by recognizing and exploiting the patterns that make them up. Sometimes the result is an explicit formula for the count; sometimes different patterns of the same arrangement will result in different expressions which are thus equal. As we shall see, many different branches of mathematics provide strikingly different ways in which we can count. This book explores some of these.

We begin our study of such patterns with some simple-looking principles – self-evident really. They hide very sophisticated and powerful enumerative techniques.

1.1 Bijections, Permutations and Sequences

There is a simple way to find whether the number of seats in a hall is the same as the number of people in it – just ask everybody to sit down. That way, we see immediately if there are any seats or people left over. This is a fundamental way of counting with a function – the function simply matches people to their seats. If there is an exact correspondence then the number of people and the number of seats are identical. Throughout this book we will effectively be doing the same thing – counting a set of objects in two different ways. The end result is an identity that expresses the cardinality of a set in two different ways. The function we use is of a very special type.

A. Camina, B. Lewis, *An Introduction to Enumeration*,
Springer Undergraduate Mathematics Series,
DOI 10.1007/978-0-85729-600-9_1, © Springer-Verlag London Limited 2011

Definition 1.1 (One–one and onto functions)

A function $f : X \to Y$ is said to be:

(i) one–one if every element $x \in X$ has a unique element $y \in Y$ such that $f(x) = y$;

(ii) onto if given any element $y \in Y$ there is an element $x \in X$ such that $f(x) = y$.

Definition 1.2 (Bijection)

A bijection between two sets is simply a one–one and onto function between the sets.

It follows that if there is a bijection between two sets then their cardinality is the same – they have precisely the same number of elements.

Definition 1.3 (Strings)

A binary string is an ordered set of digits taken from the set $\{0, 1\}$. A denary string, consists of digits taken from the set $\{0, 1, 2, \cdots, 9\}$. An alphabetic string consists of letters taken from a given alphabet. The latter are sometimes called "words".

Theorem 1.4 (Number of binary strings)

There are 2^r binary strings of length r.

Proof

There are precisely two choices for each term of the string; if the string has length r then there are precisely 2^r such strings. □

Corollary 1.5

The number of subsets of a set with r distinct elements is 2^r.

Proof

We can construct a bijection between the subsets of a set and a binary string. As we know the number of binary strings, we can count the number of subsets. Consider the set $\{1, 2, 3, \cdots, r\}$, which we can write as a set of boxes:

$$[1]\,[2]\,[3] \cdots [r]$$

We can select any subset by going along the boxes and either choosing that element, or not, as a member of a subset:

$$[\in][\notin][\in]\cdots[\in] \equiv [1][0][1]\cdots[1]$$

There are 2^r such binary strings and hence 2^r possible subsets. □

Permutations are fundamental to both group theory and enumeration, so this is an idea that we will tackle next. A permutation is no more than a function, but a function of a particular kind.

Definition 1.6 (Permutation)

A permutation of a set $X = \{x_1, x_2, \cdots, x_r\}$ is a bijection from the set X to itself.

A permutation takes a set of elements in one order and re-orders them. Not every element must have its position changed – a permutation can change no elements, one element, or even all of the elements. We can easily derive an expression for the number of permutations of r distinct objects.

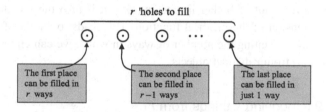

Figure 1.1 Constructing a permutation.

Theorem 1.7 (The number of permutations of r objects)

The number of permutations of r distinct objects is $r!$.

Proof

A permutation of r distinct objects assigns each object to one of the r places that make up its image (Figure 1.1). Hence the number of permutations is

$$r(r-1)\cdots 2 \cdot 1 = r!.$$

□

Note: this is an *ordered* choice – we take account of the places occupied when we make the choices. There are two ways that a permutation may be represented. We can show the effect of the permutation on each element it operates on. This is called writing the permutation in **bijective form**. We can also represent it in so-called **cycle form**. We explore these ideas in Chapter 4. For example the permutation π which operates so that $\pi(1) = 2, \pi(2) = 1, \pi(3) = 3$ and $\pi(4) = 4$ would be written in bijective form as

$$\begin{pmatrix} 1 & 2 & 3 & 4 \\ 2 & 1 & 3 & 4 \end{pmatrix}$$

and in cycle form as

$$(12)(3)(4).$$

The elements 3 and 4 are called **fixed points** because they are unchanged in the permutation. Fixed points always announce themselves as singleton cycles – that is, as a cycle with a single term.

Definition 1.8 (Zero factorial)

We define $0! = 1$.

This definition saves a lot of bother in later work – it is also the inevitable choice as the natural extension of the factorial function. This leads to another fundamental enumerative result – counting the number of ways in which we can choose k objects (without repetition) from r distinct objects.

Theorem 1.9 (Choosing k things from r)

The number of ways of choosing k objects from r is $\frac{r!}{k!(r-k)!}$, which is denoted by the Binomial coefficient $\binom{r}{k}$.

Proof

We may choose the first object in r ways, the second object in $(r-1)$ ways and the final object in $(r-k+1)$ ways. These choices may be permuted (without affecting the choice) in $k!$ ways. Hence

$$\binom{r}{k} = \frac{r(r-1)\cdots(r-k+1)}{k!} = \frac{r!}{k!(r-k)!}$$

as required.   □

Note: this also gives $\binom{0}{0} = \frac{0!}{0!0!} = 1$ exactly as required. We explore the Binomial Theorem in detail in later chapters.

Fundamental to enumeration is the idea of a sequence.

Definition 1.10 (Sequence notation)

A sequence is an ordered **list** of objects, each of which is called a **term**. Sequences are written in set notation, each term with a subscript that corresponds to its position in the list:

$$\{u_r : r \geq 0\} = \{u_0, u_1, u_2, \cdots\}.$$

For example, we can regard the number of subsets of a set with a given number of terms (Corollary 1.5) as the sequence:

$$\{s_r\} = \{1, 2, 2^2, 2^3, \cdots\}.$$

In this case we have $s_r = 2^r$. Similarly, the number of permutations of a set of objects is also enumerated by a sequence:

$$\{p_r\} = \{1, 1, 2, 6, 24, \cdots\}$$

in which $p_r = r!$. We regard all sequences as infinite – a "finite" sequence has a finite number of non-zero terms and an infinite number of zero terms.

We conclude this section by finding (with a new argument) the number of subsets a set has.

Example 1.11

Suppose that the set $\{1, 2, \cdots, r\}$ has u_r subsets. Given the new element $r+1$ we may:

(i) do nothing to each of the subsets;

(ii) add the new element $r+1$ to each subset.

In this way we construct all the subsets of the set $\{1, 2, \cdots, r+1\}$ and hence

$$u_{r+1} = 2u_r$$

and then inductively, the result follows.

Whenever we use a counting argument or insight in a proof or other work, we say that it is an **enumerative argument**. The gist of the last example was a "classic" enumerative argument. In the exercises we shall often ask for such an argument rather than other means of finding the same answer. More often than not, we find an answer by other means and then seek an enumerative way of reaching it. We usually gain a better "feel" and understanding of enumerative results in this way.

1.1.1 Exercises

Exercise 1.1

Prove by an enumerative argument that $\binom{r}{k} = \binom{r}{r-k}$.
Note: We say that the binomial coefficients are *symmetric*.

Exercise 1.2

Show by an enumerative argument that $\binom{3r}{3} = \binom{r}{1}^3 + 6\binom{r}{2}\binom{r}{1} + 3\binom{r}{3}$.
(**Hint**: $3r = r + r + r$; this is an easy result to prove algebraically and more challenging enumeratively.)

1.2 The Pigeonhole Principle

This is easily stated.

Theorem 1.12 (The pigeonhole principle)

If r sets contain $r + 1$ (or more) distinct elements then at least one of the sets contains two or more elements.

Proof

We use induction to prove it. The principle is clearly true when $r = 1$. Suppose it is true for r pigeonholes and $r + 1$ elements. If we add a new element then there are now $r + 2$ elements in all; if we add a new pigeonhole there are $r + 1$ pigeonholes in all. The new pigeonhole has:

(i) *either* 0 or 1 elements, in which case one of the other pigeonholes must contain more than 1 element – by the inductive hypothesis;

(ii) *or* 2 elements.

In either case, the induction succeeds.   □

Example 1.13

In any set of $r + 1$ integers, there is always a pair whose difference is divisible by r. There are r residue classes modulo r: two of the integers (by the pigeonhole principle) must be in the same class. Their difference is a multiple of r.

Example 1.14

A set of $r + 1$ integers is selected from the set of positive integers $\{1, 2, 3, \ldots, 2r\}$. We can show that there are always two relatively prime integers selected. Consider the "pigeonholes" which consist of pairs of consecutive integers

$$(1,2) \text{ or } (3,4) \text{ or } (5,6) \ldots \text{ or } (2r-1, 2r).$$

There are precisely r such holes. So in choosing $r + 1$ integers and then placing them in a pigeonhole, there must be a hole containing two elements. This pair is relatively prime.

Note: this example also shows the "stronger" result – in any such selection, there is always a consecutive pair selected.

1.2.1 Exercises

Exercise 1.3

You have 12 pairs of socks in a drawer, each pair distinguishable by colour. What is the most number of socks that needs to be taken from the drawer to ensure that you have a matching pair?

Exercise 1.4

How many times must you throw a dice to get a score repeated?

Exercise 1.5

An equilateral triangle has unit sides:

 (i) if five points in its interior are selected, show that there are at least two whose distance apart is less than $\frac{1}{2}$;

(ii) if ten points are selected what is the corresponding result?

Exercise 1.6

Show, using the pigeonhole principle, that $\binom{r}{k} = \frac{r(r-1)\cdots(r-k+1)}{k!}$ is a positive integer.

1.3 The Principle of Inclusion and Exclusion

We need a little work to establish this important result. First of all we give a trio of useful enumerative functions.

Definition 1.15 (Integer part functions)

For any real number x:

(i) the symbol $[x]$ denotes the integer part of x. This is called the **integer part function**;

(ii) the symbol $\lfloor x \rfloor$ denotes the largest integer less than or equal to x. This is called the **floor function**;

(iii) the symbol $\lceil x \rceil$ denotes the smallest integer greater than or equal to x. This is called the **ceiling function**.

Example 1.16

When x takes the values π and $-e$ we may find the values of $[x]$, $\lfloor x \rfloor$ and $\lceil x \rceil$:

(i) $[x] = [3.1415\ldots] = 3, \lfloor x \rfloor = 3$ and $\lceil x \rceil = 4$;

(ii) $[x] = [-2.718\ldots] = -2, \lfloor x \rfloor = -3$, and $\lceil x \rceil = -2$.

These functions frequently enable us to give remarkably compact answers to questions that appear to have different answers according to the nature of the argument involved.

Example 1.17

Suppose we are interested in the number of ways we can write the positive integer r as a sum of at most two positive integers, *without* regard to order. If r is even then $r = 2s$ and then the sums of the required form are

$$2s, \ (2s-1)+1, \ (2s-2)+2, \ \ldots, \ s+s.$$

There are $s+1$ such sums, that is

$$\frac{r}{2}+1 = \frac{r+2}{2}.$$

Now suppose that r is odd. This time we have $r = 2t+1$ and the required sums are

$$2t+1, (2t)+1, (2t-1)+2, \ldots, (t+1)+t.$$

There are $t+1$ such sums, that is

$$\frac{r-1}{2}+1 = \frac{r+1}{2}.$$

The complete answer is thus

$$\text{number of ways} = \begin{cases} \frac{r+2}{2} & \text{if } r \text{ is even;} \\ \frac{r+1}{2} & \text{if } r \text{ is odd.} \end{cases}$$

We can write this in the compact form

$$\text{number of ways} = \left[\frac{r}{2} + 1\right]$$

and it returns the correct answer whatever the parity of the argument.

Example 1.18

Suppose we wish to find the number of positive integers less than 10,000 that are divisible neither by 5 nor by 6.

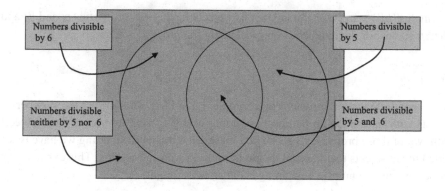

Figure 1.2   Divisibility by 5 or 6.

There are:

(i) $\left[\frac{10,000}{5}\right] = 2,000$ positive integers less than 10,000 that are divisible by 5;

(ii) and $\left[\frac{10,000}{6}\right] = 1,666$ divisible by 6.

The subtraction

$$10,000 - 2,000 - 1,666$$

does **not** give the required answer because those integers divisible by 30 $(= 5 \cdot 6)$ have been removed twice. There are $\left[\frac{10,000}{30}\right] = 333$ such integers. So the required answer we seek is

$$10,000 - 2,000 - 1,666 + 333 = 6667.$$

This example illustrates a basic enumerative technique which can be generalized to three properties (which we visualize as three mutually intersecting sets) or any number of properties. In order to state it, we need another definition.

Definition 1.19 (The counting function #)

The function # returns the number of distinct objects defined by its argument. It is also written in the form $|S|$ when it represents the cardinality of the set S.

Theorem 1.20 (Inclusion–Exclusion Principle)

Suppose that a collection of r objects have the properties α, β and γ associated with them. Then the number of objects in the collection having none of the properties α, β, γ is

$$r - (\#(\alpha) + \#(\beta) + \#(\gamma)) + (\#(\alpha\beta) + \#(\alpha\gamma) + \#(\beta\gamma)) - (\#(\alpha\beta\gamma))$$

in which $\#(\alpha\beta)$ stands for the number of objects having the properties α and β; and $\#(\alpha\beta\gamma)$ stands for the number of objects having the properties α and β and γ.

Proof

The proof follows the argument of Example 1.18. First we subtract the number of objects with each of the properties; in so doing we have twice subtracted those objects with two of those properties – so we add these back in, but in so doing we have twice added those objects that have all the properties, so we subtract this. This exhausts the properties and the number remaining is the answer sought.   □

You can see why we give it the name that it has.

Example 1.21

A positive integer r is the product of two prime factors, p_1 and p_2. We can readily find the number of positive integers less than r which are relatively prime to it. We say that any positive integer less than r has the property α_1 if it is divisible by p_1 and the property α_2 if it is divisible by p_2. So the number of positive integers less than r that are not divisible by either p_1 or p_2 is, by Theorem 1.20,

$$r - \left(\frac{r}{p_1} + \frac{r}{p_2} \right) + \frac{r}{p_1 p_2}.$$

This may be factorized:

$$r \left(1 - \frac{1}{p_1} \right) \left(1 - \frac{1}{p_2} \right)$$

We can also generalize the principle of inclusion and exclusion to any number of properties

Theorem 1.22 (The generalized principle of inclusion and exclusion)

Suppose that a collection of r objects has the properties $\alpha_1, \alpha_2, \alpha_3, \ldots$. Then the number of objects with none of these properties is

$$r - \sum_i \#(\alpha_i) + \sum_{i,j} \#(\alpha_i \alpha_j) - \sum_{i,j,k} \#(\alpha_i \alpha_j \alpha_k) + \cdots.$$

Definition 1.23 (Product notation)

The product of the terms $a_1, a_2, \ldots, a_r$ is written in the form

$$a_1 a_2 \cdots a_r = \prod_{k=1}^{r} a_k.$$

The symbol Π is the product form of the sum symbol Σ.

Example 1.24

Using this notation we may write the result of Example 1.21 in the form

$$r \prod_{\substack{p \text{ prime} \\ p \text{ divides } r}} \left(1 - \frac{1}{p}\right)$$

in which the index is descriptive rather than serial.

Next we investigate the permutations that have no fixed points.

Definition 1.25 (Derangement sequence $\{d_r\}$)

A derangement is a permutation in which no element is fixed – that is mapped to itself. The number of derangements of r distinct objects is written d_r. We define $d_0 = 1$.

Note: in terms of cycles, this means that a derangement has no singleton cycles, all its cycles are of length 2 or more (see Lemma 4.16). Think of a derangement as the number of ways of returning coursework to students so that nobody receives their own work.

Example 1.26

We can count the number of derangements in some simple examples:

(i) there is only 1 permutation of $\{1\}$ – and the (single) element is mapped to itself. So there are no derangements and hence $d_1 = 0$;

(ii) there are 2 permutations of $\{1,2\}$: $(1)(2)$ and (12). The last permutation has no fixed elements. Hence $d_2 = 1$;

(iii) there are 6 permutations of $\{1,2,3\}$:

$$(1)(2)(3), (12)(3), (13)(2), (23)(1), (123) \text{ and } (132).$$

The last 2 permutations are the only permutations that have no fixed elements and so $d_3 = 2$.

Example 1.27

We may use the principle of inclusion and exclusion to derive an explicit formula for derangements. There are $r!$ permutations of r objects. Let α_k be the property that a permutation has at least k fixed points. We may choose the fixed points from the r

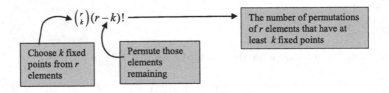

Figure 1.3 Permutations with at least k fixed points.

distinct objects in $\binom{r}{k}$ ways (Figure 1.3). The remaining $r - k$ objects can be permuted in $(r-k)!$ ways. So there are $\binom{r}{k}(r-k)!$ permutations with at least k fixed points. (Amongst the $(r-k)!$ permutations there may be some with fixed points: hence the "at least k fixed points".) By the principle of inclusion and exclusion, the number of permutations with no fixed points is

$$d_r = \binom{r}{0}r! - \binom{r}{1}(r-1)! + \binom{r}{2}(r-2)! - \cdots = \sum_{k=0}^{r}(-1)^k \binom{r}{k}(r-k)!. \quad (1.1)$$

This is the required explicit expression for derangements.

Example 1.28

We may use Equation (1.1) to evaluate d_4 rather than examining all the permutations. We have

$$d_4 = \sum_{k=0}^{4}(-1)^k \binom{4}{k}(4-k)! = 1.4! - 4.3! + 6.2! - 4.1! + 1.0!$$

$$= 24 - 24 + 12 - 4 + 1 = 9.$$

1.3.1 Exercises

Exercise 1.7

Find the number of integers less than $6,300$ that are divisible by none of 3, 4 and 5.

Exercise 1.8

How many permutations of $\{1,2,3,4\}$ are there in which:

(i) 2 is not a fixed point;

(ii) 2 and 4 are not fixed points?

Exercise 1.9

A positive integer r is the product of the square of a prime and another prime, $r = p_1^2 p_2$. How many positive integers less than r are relatively prime to it?

Exercise 1.10

Prove that for any real number x, $[x] + \left[x + \frac{1}{2}\right] = [2x]$.

1.4 The Principle of Exhaustion

Sometimes we know the total number of objects which may have multiple occurrences of a defined property. Suppose that overall, there are $f(r)$ objects of this type. We can count the objects with none of the properties, which we write #(no occurrences), those with 1 occurrence of the property, which must now be written #(1 occurrence) and so on until those with r occurrences of the property of which there are #(r occurrences). We have exhausted all the possibilities; this insight leads to another important result.

Theorem 1.29 (The principle of exhaustion)

We may write this in the form

$$f(r) = \#(\text{no occurrences}) + \#(1 \text{ occurrence}) + \cdots + \#(r \text{ occurrences}).$$

Example 1.30

We may use the principle of exhaustion to show that $2^r = \sum_{k=0}^{r} \binom{r}{k}$. There are 2^r subsets of a set with r distinct elements (Corollary 1.5). Any such subset must have *either*: no elements, and there are $\binom{r}{0}$ of these; *or* 1 element, and there are $\binom{r}{1}$ of these;

and so on *until* the subset has r elements, and there are $\binom{r}{r}$ of these. So by the principle of exhaustion:

$$2^r = \sum_{k=0}^{r} \binom{r}{k}$$

as required.

Example 1.31

Another use of the principle of exhaustion is to prove a result concerning derangement numbers. The number of permutations of r distinct objects is $r!$. We can construct and count the permutations with a given number of fixed points. Any permutation has *either* no fixed points, *or* 1 fixed point, ..., *or* k fixed points (Figure 1.4).

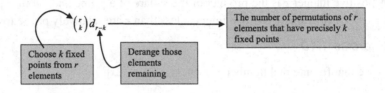

Figure 1.4 Permutations with precisely k fixed points.

By the principle of exhaustion, we must have

$$r! = \binom{r}{0} d_r + \binom{r}{1} d_{r-1} + \cdots + \binom{r}{r} d_0 = \sum_{k=0}^{r} \binom{r}{k} d_{r-k} \qquad (1.2)$$

and this is the expression sought.

1.4.1 Exercises

Exercise 1.11

Use the principle of exhaustion to prove that $\binom{r+l}{k} = \sum_{m=0}^{k} \binom{r}{m}\binom{l}{k-m}$.
(**Hint:** $r+l \rightarrow \underbrace{\circ\ \circ\cdots\circ}_{r \text{ objects}}\ \underbrace{\circ\ \circ\cdots\circ}_{l \text{ objects}}.$)

Exercise 1.12

Suppose we want to count the permutations that have no 2-cycles, a generalization of derangements. In the permutations of r objects we say there are b_r permutations with no 2-cycles (b for bicycles!). We define $b_0 = 1$.

(i) For $r = 1$, $\{1\}$ has only 1 permutation (1). So $b_1 = 1$, since it has no 2-cycles.

(ii) For $r = 2$, $\{1\,2\}$ has 2 permutations $(1)(2)$ and (12). So $b_2 = 1$.

(iii) For $r = 3$ any permutation of $\{1,2,3\}$ has *either* 0 2-cycles, *or* 1 2-cycle. Hence by the principle of exhaustion:

$$3! = b_3 + \binom{3}{1}b_1 \Rightarrow b_3 = 6 - 3.1 = 3.$$

(iv) For $r = 4$ a permutation of $\{1,2,3,4\}$ has *either* 0 2-cycles, *or* 1 2-cycle, *or* 2 2-cycles. Hence

$$4! = b_4 + \binom{4}{2}b_2 + \tfrac{1}{2!}\binom{4}{2}\binom{2}{2}b_0.$$

Use this result to find b_4 and then determine b_5.

$$2$$

Generating Functions Count

2.1 Counting – from Polynomials to Power Series

Consider the outcomes when a pair of dice are thrown and our interest is the sum of
the numbers showing. One way to model the situation is by means of a grid in which
each point (whose coordinates are non-negative integers $1 \leqslant x, y \leqslant 6$) represents one
outcome. In the grid below, the corresponding sum is shown alongside some of the
resulting points:

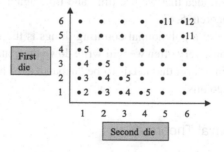

Figure 2.1 Sum of two dice.

The problem with this model is that it does not help us to portray 3 dice, 4 dice
or even more dice – the geometric insights become first difficult and then impossible.

A. Camina, B. Lewis, *An Introduction to Enumeration*,
Springer Undergraduate Mathematics Series,
DOI 10.1007/978-0-85729-600-9_2, © Springer-Verlag London Limited 2011

We need to change our model. We may represent the first die by a polynomial

$$z + z^2 + z^3 + z^4 + z^5 + z^6.$$

The symbol z here is called an **indeterminate** which means that it is not a variable that takes on values – its only role is to keep track of aspects of an enumeration. It may be replaced by X, by y or by any convenient symbol. In this instance it holds two items of information:

- the **powers** of z keep track of the different faces of the dice;

- the **coefficients** of the powers of z show the number of occurrences of each face.

The second die is represented by the same polynomial and the outcome of throwing two dice is represented – quite naturally – by

$$\left(z + z^2 + z^3 + z^4 + z^5 + z^6\right)\left(z + z^2 + z^3 + z^4 + z^5 + z^6\right).$$

By expanding this

$$z^2 + 2z^3 + 3z^4 + 4z^5 + 5z^6 + 6z^7 + 5z^8 + 4z^9 + 3z^{10} + 2z^{11} + z^{12}$$

we find that there is one way of obtaining a score of 2; there are two ways of getting a score of 3; three ways of getting a score of 4 and so on. But – and this is the crucial advantage of this algebraic model – the same method may be employed to find the number of ways that a particular score may be obtained with any number of dice. It is important to ask why this works. We can obtain a score of 9 by getting the combinations $(3,6)$, $(6,3)$, $(4,5)$ and $(5,4)$. This is precisely the same as how many ways we can get z^9 when multiplying out the product above. So the coefficients enumerate. This is an important idea that we use time and time again. We formulate this in Subsection 5.4.2 as Theorem 5.17.

This ingenious idea of a polynomial counting things is the fundamental idea behind generating functions. Very often we will write down the generating function of an enumerative example in a form that needs to be "expanded". There are two powerful tools that enable us to do this.

Theorem 2.1 (Binomial Theorem)

For a given integer r

$$(1+z)^r = \begin{cases} \sum\limits_{k \geqslant 0} \binom{r}{k} z^k & \text{if } r > 0; \\ 1 & \text{if } r = 0; \\ \sum\limits_{k \geqslant 0} (-1)^k \left(\binom{-r+k-1}{k}\right) z^k & \text{if } r < 0. \end{cases}$$

Moreover, the Binomial coefficients involved are given by the explicit form

$$\binom{r}{k} = \frac{r(r-1)\cdots(r-k+1)}{k!} = \frac{r!}{k!(r-k)!}.$$

Convention: the sum $\sum_{k=0}^{r} \binom{r}{k} z^k$ may be written $\sum_{k\geqslant 0} \binom{r}{k} z^k$ in which the summation index k takes the values $k = 0$ to $k = \infty$. As soon as $k > r$ the Binomial coefficients are zero. This convention leads to concise ways of writing and dealing with such sums.
Convergence: the first form of the sum is a finite sum because $\binom{r}{k} = 0$ if $k > r$. It is therefore valid for all values of z. This second form is valid for all values of z except -1, since 0^0 is undefined. The third form is a non-terminating expression (so it is an infinite expansion); it is only valid when $|z| < 1$.

Example 2.2

The Binomial Theorem makes it easy to expand powers of expressions. For example:

(i)

$$(2+3z)^7 = 2^7 \left(1 + \frac{3z}{2}\right)^7 = 2^7 \sum_{r\geqslant 0} \binom{7}{r}\left(\frac{3z}{2}\right)^r$$

$$= 2^7 + 2^7 \binom{7}{1}\frac{3z}{2} + 2^7 \binom{7}{2}\frac{3^2 z^2}{2^2} + \cdots + 2^7 \binom{7}{7}\frac{3^7 z^7}{2^7}$$

$$= 128 + 1344z + 6048z^2 + \cdots + 2187z^7;$$

(ii)

$$(2+3z)^{-7} = 2^{-7} \left(1 + \frac{3z}{2}\right)^{-7} = 2^{-7} \sum_{r\geqslant 0} \binom{7+r-1}{r}\left(\frac{3z}{2}\right)^r$$

$$= 2^{-7} + 2^{-7} \binom{7}{1}\frac{3z}{2} + 2^{-7} \binom{8}{2}\frac{3^2 z^2}{2^2} + \cdots$$

$$= \frac{1}{128} + \frac{21}{256}z + \frac{63}{128}z^2 + \cdots.$$

This expansion is valid only when $\left|\frac{3z}{2}\right| < 1$ that is, when $|z| < \frac{2}{3}$.

Theorem 2.3 (The sum of a geometric progression (GP))

In the finite case, we have:

$$1 + z + z^2 + \cdots + z^n = \sum_{r=0}^{n} z^r = \begin{cases} \frac{1-z^{n+1}}{1-z} & \text{if } z \neq 1; \\ n+1 & \text{otherwise.} \end{cases} \tag{2.1}$$

If $|z| < 1$ we can evaluate the infinite sum:

$$1 + z + z^2 + \cdots = \sum_{r \geqslant 0} z^r = \frac{1}{1-z}. \tag{2.2}$$

We return to generating functions and the way in which these results may be exploited.

Example 2.4

We explore the number of ways there are to obtain a score of 12 with the throw of five dice. Simple: it is the coefficient of z^{12} in the product of five polynomials, each of which enumerates the outcomes of a single die:

$$\left(z + z^2 + z^3 + z^4 + z^5 + z^6 \right)^5.$$

Now we proceed to expand and simplify this using Theorem 2.1 and Equation (2.1). We have,

$$\left(z + z^2 + z^3 + z^4 + z^5 + z^6 \right)^5 = z^5 \left(1 + z + z^2 + z^3 + z^4 + z^5 \right)^5$$

$$= \frac{z^5 \left(1 - z^6 \right)^5}{(1-z)^5}$$

$$= z^5 \left(1 - z^6 \right)^5 (1-z)^{-5}$$

$$= \left(z^5 - 5z^{11} + \cdots - z^{35} \right) \sum_{r \geqslant 0} \binom{r+4}{4} z^r.$$

So the coefficient of z^{12} is just

$$\binom{7+4}{4} - 5\binom{1+4}{4} = 330 - 25 = 305.$$

There are 305 ways to obtain a score of 12 when five dice are thrown.

Example 2.5

A generous father wishes to divide £20 between his daughters Emma and Pippa, and son Leon, so that they each receive at least £5, nobody receives more than £10, and Emma gets an even amount. How many ways are there of doing this? What we seek are the non-negative, integer solutions to the equation $e + p + l = 20$ subject to the conditions that $5 \leqslant e, p, l \leqslant 10$ and e is even. Viewed in this way, we may associate a polynomial with each recipient:

$$z^5 + z^6 + z^7 + z^8 + z^9 + z^{10}$$

for Pippa and Leon, together with

$$z^6 + z^8 + z^{10}$$

for Emma. Each of these enumerates the amounts they might receive. The answer to our problem is the coefficient of z^{20} in the product

$$\left(z^6 + z^8 + z^{10}\right)\left(z^5 + z^6 + z^7 + z^8 + z^9 + z^{10}\right)^2$$

Once again, thanks to Theorem 2.1 and Equation (2.1) it is easier to find than it looks. The expression may be re-written, and then manipulated

$$\left(z^6 + z^8 + z^{10}\right)\left(z^5 + z^6 + z^7 + z^8 + z^9 + z^{10}\right)^2 = z^{16}\left(1 + z^2 + z^4\right)\left(\frac{1 - z^6}{1 - z}\right)^2$$

$$= z^{16}\frac{\left(1 + z^2 + z^4\right)\left(1 - 2z^6 + z^{12}\right)}{(1 - z)^2}$$

$$= \left(z^{16} + z^{18} + \cdots + z^{32}\right)\sum_{r \geqslant 0}\binom{r+1}{1}z^r.$$

The required coefficient of z^{20} is now easy to pick out. It is

$$\binom{5}{1} + \binom{3}{1} + \binom{1}{1} = 5 + 3 + 1 = 9.$$

The expressions we have used to help us count in Examples 2.4 and 2.5 were made up from polynomials. This is because the terms in which we were interested only had a finite number of possibilities. In Example 2.4 the only possible scores are $5, 6, 7, \ldots, 28, 29, 30$. They make up a finite sequence $\{1, 2, \ldots, 30\}$.

Many enumerations are not finite and result in a sequence that does not terminate: for example the number of subsets of a set with r elements. In dealing with an infinite sequence, we need infinite expressions. This leads us to the idea of a **generating function**.

Definition 2.6 (Generating function)

Given any sequence $\{u_r\} = \{u_0, u_1, u_2, \cdots\}$, a generating function $U(z)$ for the sequence is an expression (called a **power series** in the infinite case and a polynomial in the finite case) in which,

$$U(z) = u_0 + u_1 z + u_2 z^2 + \cdots = \sum_{r \geqslant 0} u_r z^r.$$

This definition has two parts:

(i) the **power** series (or polynomial) on the right, each of whose coefficients is a term of the sequence placed against a power of the indeterminate z that matches its position in the sequence;

(ii) the **function** $U(z)$ that explicitly represents the power series, or polynomial following some re-arrangement.

If we can find the function $U(z)$, that is manipulate the power series into a new, simpler form using tools such as the Binomial Theorem and sums of a GP, then we may employ other results from analysis on it: in so doing, we unearth information about the sequence itself. Think of a generating function as a reel of magnetic tape, divided up into a number of segments – possibly infinite. Each segment is coded with a term of the sequence, which is also the coefficient of the corresponding power of z; each power of z is simply the address of an individual segment.

Figure 2.2   Generating function as a reel of magnetic tape.

The name for the reel of magnetic tape is the new form of the function. The importance of a generating function is that it is also "hard wired" with many other important aspects of the sequence. In ways we will explore and discover, the generating function $U(z)$ encapsulates crucial information about the sequence $\{u_r\}$. Problems of counting – enumeration – are the inspiration for generating functions, and they provide a powerful, versatile and robust tool to solve and explore these problems.

$$P(z) = (1 + z + z^{1+1} + z^{1+1+1} + \cdots)(1 + z^2 + z^{2+2} + z^{2+2+2} + \cdots)(1 + z^4 + z^{4+4} + z^{4+4+4} + \cdots)$$

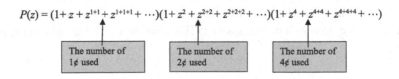

Figure 2.3   Giving change for 100¢.

Example 2.7

Imagine a country in which there are only three coins: a 1¢ coin, a 2¢ coin and a 4¢ coin – this is a very mathematical country. We determine the generating function for the enumeration of the number of ways that change may be given for 100¢. We need to solve the equation $r + 2s + 4t = 100$ for non-negative integers r, s and t. We do this by finding the coefficient of z^{100} in a generating function (Figure 2.3).

We could, of course, multiply out initial terms of each bracket, and with some perseverance we might eventually find the coefficient of z^{100}. But now we exploit the generating function by concentrating on the right-hand side – seeking an easier form for it.

We start by writing the generating function in the form

$$P(z) = (1 + z + z^2 + z^3 + \cdots)(1 + z^2 + z^4 + z^6 + \cdots)(1 + z^4 + z^8 + z^{12} + \cdots).$$

And now for the decisive step: each of the bracketed expressions is a GP, so when $|z| < 1$, we have

$$P(z) = \frac{1}{(1 - z)(1 - z^2)(1 - z^4)}.$$

Now we use partial fractions and find that

$$P(z) = \frac{1}{8(1 - z)^3} + \frac{1}{4(1 - z)^2} + \frac{9}{32(1 - z)} + \frac{1}{16(1 + z)^2} + \frac{5}{32(1 + z)} + \frac{1 + z}{8(1 + z^2)}.$$

We may expand each term on the right using the Binomial Theorem; the coefficient of z^{100} is simply:

$$\frac{1}{8}\binom{102}{2} + \frac{1}{4}\binom{102}{1} + \frac{9}{32} + \frac{(-1)^{100}}{16}\binom{101}{1} + \frac{5(-1)^{100}}{32} + \frac{(-1)^{50}}{8} = 676.$$

That's the power of generating functions – it even provides the answer for any amount of money, but we'll pick that up again later. This example made use of another powerful tool: **partial fractions**. This is a technique that comes into its own with generating functions.

2.1.1 Exercises

Exercise 2.1

What is the generating function for the score obtained when ten dice are thrown? Use this to find the number of ways that a score of 25 may be obtained. (Leave your answer in binomial form.)

Exercise 2.2

The equation $x + 2y + 4z = 100$ determines a plane in three-dimensional Euclidean space. How many non-negative lattice points (a lattice point is a point each of whose coordinates is an integer) lie on this plane?

Exercise 2.3

How many ways are there to give change for £2 if the coinage is 1p and 3p.

Exercise 2.4

Show that the distinct divisors of $p_1^2 p_2^3$ (where p_1 and p_2 are primes) are generated by the expression

$$\left(1 + p_1 + p_1^2\right)\left(1 + p_2 + p_2^2 + p_2^3\right).$$

Deduce that an arbitrary positive integer r whose prime factorization is

$$r = p_1^{a_1} p_2^{a_2} \cdots p_k^{a_k}$$

has

$$\prod_{m=1}^{k} (1 + a_m)$$

distinct divisors.

2.2 Recurrence Relations and Enumeration

When we examine a particular enumeration, we frequently resort to breaking down one of its configurations into smaller parts, so that we can understand how it is made up. Mathematically this may be expressed as a relation between configurations of different "sizes". Such an expression is called a **recurrence relation** and they come in all sorts of varieties.

Definition 2.8 (Recurrence relation (informal))

The recurrence relation satisfied by a sequence is a recipe that uses initial terms as the ingredients for subsequent terms. It is usually "seeded" by some initial values.

Although recurrences are often the starting point in the analysis of an enumeration, deriving a recurrence relation is by no means easy or obvious. We have a number of tools in the enumerative toolbox that help, such as the principle of exhaustion, that were set out in Chapter 1. This section concentrates on this crucial first step – constructing a recurrence by analyzing how the objects of different sizes fit together.

Example 2.9

We start by finding a recurrence satisfied by the number of denary strings of length r, see Definition 1.3, in which between them the digits $\{3,6,9\}$ occur an even number of times. The strategy is to construct an object of "size" $r+1$ from those of "size" r.

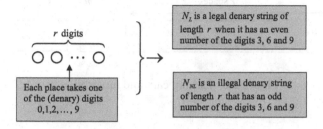

Figure 2.4 Denary strings.

Call the number of legal strings of length r, n_r. A legal string of length $r+1$ has one of the forms $N_L a$ or $N_{NL} b$ where:

– N_L is a legal string of length r so $a \in \{0,1,2,4,5,7,8\}$;

– N_{NL} is not a legal string of length r so $b \in \{3,6,9\}$.

It follows that

$$n_{r+1} = \#(a).\#(N_L) + \#(b).\#(N_{NL}) = 7\#(N_L) + 3\#(N_{NL})$$
$$= 3[\#(N_L) + \#(N_{NL})] + 4\#(N_L).$$

The total number of strings of length r is 10^r and so

$$n_{r+1} = 3.10^r + 4\#(N_L) = 3.10^r + 4n_r.$$

This is the required recurrence. Once we know one initial value we can use the recurrence to creep forwards, finding successive terms of the sequence, one at a time.

Example 2.10

We have $n_1 = 7$ so that $n_2 = 3.10^1 + 4n_1 = 58$ and $n_3 = 3.10^2 + 4n_2 = 532$; we may continue this as far as we please.

Here is another example, this time with a geometric flavour – and with a sequence as answer that turns out to be very important.

Example 2.11

A car park consists of a row of r spaces; a motorbike (m) takes one space, a car (c) two.

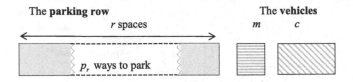

Figure 2.5 Parking cars and motorbikes.

We seek the number of ways there are of filling the spaces. Call the number of ways of filling r spaces, p_r. The strategy is to use the principle of exhaustion on the final parking space. *Either* there is a car *or* a motorbike in the final space and so the form of any parking arrangement has one of the two possibilities $P_{r-1}m$ or $P_{r-2}c$ where

– P_{r-1} is a parking arrangement of $r-1$ places, and m is a motorbike;

– P_{r-2} is a parking arrangement of $r-2$ places and c is a car.

Hence,

$$\#(P_r) = \#(m).\#(P_{r-1}) + \#(c).\#(P_{r-2})$$

which means that the required recurrence is

$$p_r = p_{r-1} + p_{r-2}.$$

We easily find that $p_1 = 1$ and $p_2 = 2$, so then $p_3 = 3, p_4 = 5, p_5 = 8$, and the sequence proceeds, each term being the sum of the two preceding terms.

The terms of this sequence are called **Fibonacci numbers** and we meet them on many occasions. We define $p_0 = 1$ as this is the only value consistent with the recurrence relation. The sequence $\{p_r\}$ is *not* the Fibonacci sequence, which we denote by $\{F_r\}$. In fact we have

$$p_r = F_{r+1}$$

and so the sequence $\{p_r\}$ *does* consist of Fibonacci numbers, but they are displaced by one step.

Example 2.12

Next we explore another situation that also leads to Fibonacci numbers, again displaced. We seek to count the number of ways there are to select subsets (including the empty set) from the set $\{1,2,3,\ldots,r\}$ in which no consecutive numbers occur. Suppose the number of ways is s_r. Then in selecting such a subset it *either* includes the element r *or not* (Figure 2.6).

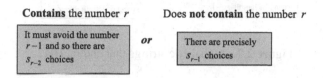

Figure 2.6 Subsets without consecutive terms.

Overall then, the number of subsets s_r is

$$s_r = s_{r-1} + s_{r-2}.$$

However, for the set $\{1\}$ there are two subsets, $\emptyset$ and $\{1\}$, neither of which contains consecutive numbers. So $s_1 = 2$. Similarly, $s_2 = 3$. If we want the recurrence to be true for $r = 0$ then we must set $s_0 = 1$. The sequence continues (each term being the sum of the two previous terms) $\{s_r\} = \{1,2,3,5,8,\ldots\}$ and each term is also a Fibonacci number, this time displaced two places:

$$s_r = F_{r+2}.$$

Time to define the Fibonacci sequence.

Definition 2.13 (Fibonacci sequence)

The Fibonacci sequence $\{F_r\}$ satisfies the recurrence $F_r = F_{r-1} + F_{r-2}$ with the initial terms $F_0 = 0$ and $F_1 = 1$.

The Fibonacci sequence has the initial terms $\{F_r\} = \{0,1,1,2,3,5,8,\ldots\}$.

Example 2.14

This time we count the number of ways there are to toss a coin until it ends (for the first time) on double heads. Suppose that there are t_r ways to do this. So we require the number of alphabetic strings of length r in which the final two places are H followed by another H, preceded by a string of Hs and Ts in which the string "HH" does not occur. Such a string **starts** *either* with a tail *or* a head (which is then followed by a tail): Figure 2.7.

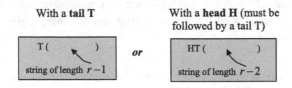

With a **tail T**

With a **head H** (must be followed by a tail T)

T (   )
string of length $r-1$

or

HT (   )
string of length $r-2$

Figure 2.7 Alphabetic strings that end on HH.

Overall then, the number of strings t_r is

$$t_r = t_{r-1} + t_{r-2}.$$

For one toss of a coin the outcomes are either T or H. So $t_1 = 0$ since neither of these ends in a double head. Similarly, $t_2 = 1$, $t_3 = 1$ and hence

$$t_r = F_{r-1}.$$

If we define $F_{-1} = 1$ (a value consistent with the Fibonacci recurrence relation) then this is true when $r \geqslant 0$.

The next example exploits Fibonacci numbers but results in another sequence of importance.

Example 2.15

We count the number of ways we can select a subset from the **circular** set $\{1, 2, 3, \ldots, r\}$ that does not contain consecutive numbers. (Circular here, means that we regard 1 and r as consecutive, and again, we include the empty set as an allowable subset choice.) Suppose the number of such selections is L_r. A selection *either* contains the element r, *or not* (Figure 2.8).

From Example 2.12 we know the number of ways that subsets (without consecutive elements) may be chosen from a set of r elements is F_{r+2}. Overall then, we have

$$L_r = F_{r-3+2} + F_{r-1+2} = F_{r-1} + F_{r+1}.$$

The sequence that appears here is a sibling sequence to the Fibonacci sequence.

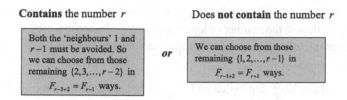

Contains the number r

Both the 'neighbours' 1 and $r-1$ must be avoided. So we can choose from those remaining $\{2,3,\ldots,r-2\}$ in $F_{r-3+2} = F_{r-1}$ ways.

or

Does not contain the number r

We can choose from those remaining $\{1,2,\ldots,r-1\}$ in $F_{r-1+2} = F_{r+1}$ ways.

Figure 2.8 Subsets of a circular set.

Definition 2.16 (Lucas sequence)

The sequence $\{L_r\}$, defined as the sum of pairs of Fibonacci numbers:

$$L_r = F_{r-1} + F_{r+1}$$

is called the Lucas sequence, and its terms are Lucas numbers. It has the initial terms $\{L_r\} = \{2,1,3,4,7,11,18,\ldots\}$ and satisfies the recurrence

$$L_r = L_{r-1} + L_{r-2}.$$

Note: the Lucas and Fibonacci sequences share the same recurrence relation (you will be asked to prove this in the exercises). They are distinguished by the two terms with which they begin: $\{0,1\}$ for the Fibonacci sequence and $\{2,1\}$ for the Lucas sequence. The same is true for the sequences of Example 2.11, $\{p_r\}$, and of Example 2.12, $\{s_r\}$: they too obey the same recurrence – they are distinguished by their two initial terms.

Example 2.17

Suppose there is an inexhaustible number of coloured counters – red, yellow, green and blue. We seek a recurrence for the number of ways that r of them can be stacked so that there are no adjacent red counters (Figure 2.9).

A legal stack has no adjacent red counters.

Figure 2.9 No adjacent red counters.

The enumerative strategy is to construct a stack of "size" $r+1$ from one of "size" r. Suppose that we denote the number of legal stacks by s_r. In constructing s_{r+1} for $r \geqslant 2$ we use the principle of exhaustion on the topmost counter; every legal stack *either* has the form $S_R c_1$ *or* $S_{NR} c_2$, where:

(i) S_R is a legal stack (of size r) that ends with a red counter, so that $c_1 \in \{Y, G, B\}$;

(ii) S_{NR} is a legal stack (of size r) that does not end with a red counter, so that $c_2 \in \{R, Y, G, B\}$.

It follows that,

$$s_{r+1} = 3\#(S_R) + 4\#(S_{NR})$$
$$= 3\left[\#(S_R) + \#(S_{NR})\right] + \#(S_{NR}) = 3s_r + \#(S_{NR}).$$

However $\#(S_{NR}) = 3s_{r-1}$ since it can be made from a legal stack of size $r-1$ followed by a final counter that may be chosen from three pieces: $\{Y, G, B\}$. So

$$s_{r+1} = 3s_r + 3s_{r-1}.$$

We note that $s_1 = 4$, $s_2 = 15 = (4.4 - 1)$ so that $s_3 = 3.s_2 + 3.s_1 = 57$ and $s_4 = 3.57 + 3.15 = 216$. Again, we can continue calculating as many terms of the sequence as desired – though it does become very tedious.

Example 2.18

A geometric example: finding the number of regions created when r lines are drawn in the plane – no two of which are parallel, and no three of which are coincident (Figure 2.10).

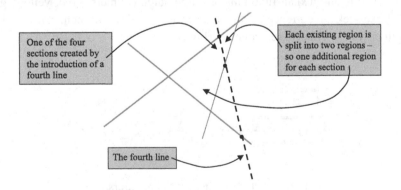

Figure 2.10 Regions created by four lines.

Suppose we denote the number of regions by R_r. The rth line meets each of the other $r-1$ lines in a distinct point (because of the conditions) and these $r-1$ points of intersection divide the rth line into r sections. Each of these sections divides an existing region into two regions, adding one new region.

So, when the rth line is drawn, the number of regions increases by r. We have the recurrence:

$$R_r = R_{r-1} + r \text{ and } R_0 = 1.$$

We met the derangement sequence $\{d_r\}$ in Equation (1.2); there we showed that its terms, called derangement numbers, satisfy a rather complicated recurrence relation:

$$r! = \sum_{k=0}^{r} \binom{r}{k} d_{r-k}.$$

In fact, it also satisfies a much simpler recurrence relation. This next example shows how this may be constructed enumeratively.

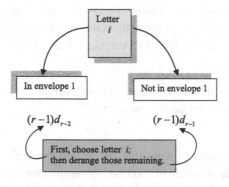

Figure 2.11 Where does letter i go?

Example 2.19

The recurrence relation is

$$d_r = (r-1)d_{r-2} + (r-1)d_{r-1}.$$

Suppose that we have a set of r letters $l_1, l_2, \ldots, l_r$ and r envelopes correctly (and distinctly) addressed $e_1, e_2, \ldots, e_r$. We seek the number of derangements in which no letter l_i goes into the correct envelope e_i. We call this number d_r. In any of these derangements letter 1 must be placed in envelope i where $2 \leqslant i \leqslant r$. Now we exhaust the possibilities for letter i (see Figure 2.11):

– *either* letter i is in envelope 1

– *or* letter i is not in envelope 1.

Overall then, we have

$$d_r = (r-1)d_{r-2} + (r-1)d_{r-1}.$$

The importance of this new recurrence is its simplicity. It may also be derived directly from the explicit formula (Equation 1.1, page 12) – see Exercise 2.8.

2.2.1 Exercises

Exercise 2.5

Let a_r be the number of legal arithmetic "expressions" that are made up of r items chosen from the operations $\{+, \times, /\}$ and the digits $\{0, 1, 2, \ldots, 9\}$. A legal arithmetic expression is one that can be evaluated: $2 + 3 \times 5$ is legal, as is 57; while $8 + \times 9$ is not. Find a recurrence relation for a_r.

Exercise 2.6

Find a recurrence for the number of binary strings, made up of r digits drawn from $0, 1$ which do not have consecutive 0s.

Exercise 2.7

Let u_r be the number of ways that the natural number r may be written as a sum of 1s and 2s, in which the order of the summands is counted. For example: $u_2 = 2$ since $2 \equiv 2 = 1 + 1$, while $u_3 = 3$ since $3 \equiv 2 + 1 = 1 + 2 = 1 + 1 + 1$. Find a recurrence relation satisfied by the sequence u_r.

Exercise 2.8

Given the explicit formula (Equation 1.1)

$$d_r = \sum_{k=0}^{r} (-1)^k \binom{r}{k} (r-k)!$$

prove that the derangement sequence satisfies the recurrence $d_r = rd_{r-1} + (-1)^r$. Then show that this recurrence leads to the recurrence of Example 2.19.

Exercise 2.9

Prove that the Lucas sequence satisfies the same recurrence as that for the Fibonacci sequence:

$$L_r = L_{r-1} + L_{r-2}.$$

2.3 Sequence to Generating Function

There are two routes to a generating function. The first starts with the sequence itself, and assumes that we know all of its terms.

2.3.1 Sequence to Generating Function

We start with a simple example of a sequence with known terms and seek to find its generating function – that is, we want this as a function in an explicit form rather than as a power series.

Example 2.20

We can easily find the generating function of the sequence $\{u_r\} = \{1,1,1,\ldots\}$. The generating function is the power series

$$1 + 1z + 1z^2 + \cdots = 1 + z + z^2 + \cdots.$$

However, we note that if $|z| < 1$ this is a convergent geometric progression, so by Equation (2.2)

$$U(z) = \frac{1}{1-z}.$$

The function $U(z)$ is the generating function for the sequence $\{u_r\} = \{1,1,1,\ldots\}$. We may use this generating function to find generating functions for other sequences.

Example 2.21

The generating function of the last example may be written like this

$$\frac{1}{1-z} = 1 + z + z^2 + z^3 + \cdots.$$

If we differentiate both sides then we obtain:

$$\frac{1}{(1-z)^2} = 1 + 2z + 3z^2 + 4z^3 + \cdots.$$

(Term-by-term differentiation of a power series is valid when it is absolutely convergent; in this case when $|z| < 1$.) This new expression is simply another generating function for another sequence: the function

$$\frac{1}{(1-z)^2}$$

is the generating function for the sequence

$$\{1,2,3,\ldots\} = \{r+1\}.$$

We can also integrate absolutely convergent power series to create another "new sequence for old". Using operations like this we can build a library of sequences and their generating functions. There are other operations we can use as well.

Example 2.22

If we multiply the generating function of the last example by z we find that

$$\frac{z}{(1-z)^2} = z + 2z^2 + 3z^3 + 4z^4 + \cdots.$$

We may conclude that the sequence

$$\{0,1,2,3,\ldots\} = \{r\}$$

has the generating function

$$\frac{z}{(1-z)^2}.$$

So our generating function library now consists of

Sequence	Generating Function
$\{1,1,1,\ldots\}$	$\dfrac{1}{1-z}$
$\{r+1\} = \{1,2,3,\ldots\}$	$\dfrac{1}{(1-z)^2}$
$\{r\} = \{0,1,2,3,\ldots\}$	$\dfrac{z}{(1-z)^2}$

Table 2.1 Generating function library.

We will build the library as the book progresses and it is summarized in Appendix A.

2.3.2 Recurrence to Generating Function

Most enumerations start with a recurrence relation that has been derived, rather than the enumerative sequence itself. Once we have such a recurrence relation for a sequence of interest, we have seen that it can be used to find successive terms of the sequence. But there are much more powerful ways of exploiting the recurrence to find this and other properties of the sequence – the route to these methods goes through a generating function. So we turn to the problem of converting a recurrence into a generating function. There is a standard way that we go about this task – and it follows a recipe with three steps.

Algorithm 2.23 (Recurrence to generating function – three-step recipe)

The three parts of the recipe to convert a recurrence into a generating function are:

 (i) write out recurrence;

(ii) multiply through by z^r and sum for valid r;

(iii) invoke generating and other library functions.

Example 2.24

We illustrate the recipe by converting the recurrence $u_r = 5u_{r-1} - 6u_{r-2}$ with the initial values $u_0 = 5$ and $u_1 = 12$ for the sequence $\{u_r\}$ into a generating function.

 (i) The first step is $u_r = 5u_{r-1} - 6u_{r-2}$.

(ii) The second step then becomes,

$$\sum_{r \geqslant 2} u_r z^r = \sum_{r \geqslant 2} 5u_{r-1} z^r - \sum_{r \geqslant 2} 6u_{r-2} z^r. \tag{2.3}$$

   (Note that as there is a term $r - 2$ in the recurrence, we take $r \geqslant 2$.)

(iii) The final step is where all the action takes place. The generating function we seek is, $U(z) = \sum_{r \geqslant 0} u_r z^r$ which is almost the term on the left of Equation (2.3); so we write,

$$U(z) = u_0 + u_1 z + \sum_{r \geqslant 2} u_r z^r = 5 + 12z + \sum_{r \geqslant 2} u_r z^r$$

$$\Rightarrow \sum_{r \geqslant 2} u_r z^r = U(z) - 5 - 12z.$$

Now for the first term on the right of Equation (2.3). We have,

$$\sum_{r \geqslant 2} 5u_{r-1} z^r = 5z \sum_{r \geqslant 2} u_{r-1} z^{r-1} = 5z \left(\sum_{r \geqslant 1} u_{r-1} z^{r-1} - u_0 \right)$$

$$= 5zU(z) - 25z.$$

The final term is now easy

$$\sum_{r \geqslant 2} 6u_{r-2}z^r = 6z^2 \sum_{r \geqslant 2} u_{r-2}z^{r-2} = 6z^2 U(z).$$

We put each of these results into Equation (2.3):

$$U(z) - 5 - 12z = 5zU(z) - 25z - 6z^2 U(z)$$

and then solve this for $U(z)$. We find that

$$U(z) = \frac{5 - 13z}{1 - 5z + 6z^2}.$$

We have quickly passed over a very important idea which we now make explicit. This important result, whose proof is immediate, should become second nature.

Theorem 2.25 (Re-indexing a sum)

We have

$$U(z) = \sum_{r \geqslant 0} u_r z^r = \sum_{r \geqslant 1} u_{r-1}z^{r-1} = \sum_{r \geqslant 2} u_{r-2}z^{r-2} \ldots$$

Most of the recurrences that we encountered in the last section can be converted into generating functions by the three-step recipe. A notable exception is the derangement sequence – we must wait for Chapter 7 for that.

Example 2.26

The recurrence relation satisfied by the Fibonacci sequence $\{F_r\}$ is $F_r = F_{r-1} + F_{r-2}$ with the initial values $F_0 = 0$ and $F_1 = 1$. Applying the three-step recipe gives:

(i) The recurrence is $F_r = F_{r-1} + F_{r-2}$.

(ii) Then we sum on the index r with corresponding powers of z:

$$\sum_{r \geqslant 2} F_r z^r = \sum_{r \geqslant 2} F_{r-1}z^r + \sum_{r \geqslant 2} F_{r-2}z^r.$$

(iii) Finally

$$\sum_{r \geqslant 0} F_r z^r - F_0 - F_1 z = z \left(\sum_{r \geqslant 0} F_{r-1}z^{r-1} - F_0 \right) + z^2 \sum_{r \geqslant 0} F_r z^r.$$

Using the initial values and denoting the generating function by $F(z)$, we have

$$F(z) - 0 - z = z(F(z) - 0) + z^2 F(z)$$

and we can solve this for the required generating function $F(z)$:

$$F(z) = \sum_{r \geqslant 0} F_r z^r = \frac{z}{1 - z - z^2}.$$

Working in the same way with the recurrence $L_r = L_{r-1} + L_{r-2}$ with the initial values $L_0 = 2$ and $L_1 = 1$ of the Lucas sequence and denoting the generating function by $L(z)$, we find that

$$L(z) = \sum_{r \geqslant 0} L_r z^r = \frac{2 - z}{1 - z - z^2}.$$

Note: the generating functions of these two sequences have the same denominator; also note that they obey the same recurrence. We will explore this later.

Example 2.27

We will convert the recurrence of Example 2.18 into a generating function.

(i) The recurrence is:

$$R_r = R_{r-1} + r \text{ and } R_0 = 1.$$

(ii) The next step is

$$\sum_{r \geqslant 1} R_r z^r = \sum_{r \geqslant 1} R_{r-1} z^r + \sum_{r \geqslant 1} r z^r.$$

(iii) In this final step we draw on our library of generating functions for the last term on the right. We can also re-write the other terms. We then have:

$$\sum_{r \geqslant 0} R_r z^r - R_0 = z \sum_{r \geqslant 0} R_r z^r + \frac{z}{(1 - z)^2}.$$

If we let the generating function be $R(z)$ then we have

$$R(z) - 1 = zR(z) + \frac{z}{(1 - z)^2}$$

and hence

$$R(z) = \frac{1 - z + z^2}{(1 - z)^3}.$$

2.3.3 Exercises

Exercise 2.10

Find generating functions for the sequences

(i) $\{v_r\} = \{1, -1, 1, -1, \ldots\}$;

(ii) $\{u_r\} = \{1, -2, 3, -4, \ldots\}$;

(iii) $\{o_r\} = \{1, 0, 3, 0, 5, 0, \ldots\}$.

Exercise 2.11

Integrate the expression $\frac{1}{1-z} = 1 + z + z^2 + \cdots$ and determine the constant of integration by assigning the value $z = 0$. What is the generating function for the Reciprocal sequence, $\{0, 1, \frac{1}{2}, \frac{1}{3}, \frac{1}{4}, \ldots\}$?

Exercise 2.12

A sequence $\{u_r\}$ obeys the recurrence relation, $u_r = u_{r-1} + 2u_{r-2}$ with $u_0 = 4$ and $u_1 = 5$. Find the generating function for the sequence.

Exercise 2.13

The sequence $\{a_r\}$ satisfies the recurrence relation $a_r = 2a_{r-1} + 15a_{r-2}$ with $a_0 = 4$ and $a_1 = 4$. Find the generating function of the sequence $\{a_r\}$.

Exercise 2.14

Use the recurrence and the initial terms of the Lucas sequence $\{L_r\}$ to confirm that its generating function is as given in Example 2.26.

2.4 Miscellaneous Exercises

Exercise 2.15

Find a generating function for the number of strings of length r made up from the digits $\{0, 1, 2, 3\}$ in which there is never a 3 anywhere to the right of 0.

Exercise 2.16

We define the matrix M as the 2×2 array:

$$M = \begin{pmatrix} 0 & 1 \\ 1 & 1 \end{pmatrix}.$$

The powers of this matrix have a number of surprising connections with Fibonacci and Lucas numbers.

(i) Show that

$$M^r = \begin{pmatrix} F_{r-1} & F_r \\ F_r & F_{r+1} \end{pmatrix}.$$

(ii) show that the trace of M^r is given by $\operatorname{tr}(M^r) = L_r$;

(iii) by considering the determinant of M^r prove **Cassini's identity**

$$F_{r-1}F_{r+1} - F_r^2 = (-1)^r.$$

Exercise 2.17

Find a generating function for the number of non-negative integer solutions to the equation $a + 2b = r$ for each positive integer r.

Exercise 2.18

A mathematics examination consists of six modules each with m marks. Show that the number of ways a candidate may score precisely $3m$ marks (the "pass mark") overall is

$$\binom{3m+5}{5} - 6\binom{2m+4}{5} + 15\binom{m+3}{5}.$$

Exercise 2.19

In this exercise we investigate the number of ways p_r in which a positive integer may be written as the **ordered** sums of the summands $\{1, 2, 3\}$. For example: one can only be written as 1 so $p_1 = 1$; however two may be written as $2 = 2 = 1 + 1$ and hence $p_2 = 2$. Three has four ways of being written, $3 = 3 = 2 + 1 = 1 + 2 = 1 + 1 + 1$ and hence $p_3 = 4$. Find a recurrence relation for the terms of the sequence $\{p_r\}$.

Exercise 2.20

You are given two tiles – one a unit square, and the other a rectangle made up from two unit squares. The rectangle can be laid vertically or horizontally. We denote the number of ways of tiling a $2 \times r$ rectangle by f_r. Show that $f_r = 2f_{r-1} + 3f_{r-2}$ and hence find a generating function for the number of tilings.

3

Working with Generating Functions

We have explored a range of enumerative problems by finding a recurrence relation that the corresponding count satisfies. This means that once we have some initial terms, we can progressively calculate more. We have also developed a way to convert some recurrence relations into a generating function. If we can manipulate the generating function and then expand it as a power series that enables us to give explicit expressions for the count by reading off the coefficient of a particular power of the indeterminate used. This last step is the focus of this chapter. Along the way we also explore the different recurrences satisfied by the corresponding sequence.

3.1 Expanding Generating Functions

We start with a generating function of a particularly form.

Example 3.1 (Generating functions with a simple denominator)

We start with the sequence that counted the number of regions in the plane created by intersecting lines – Example 2.18. The generating function is

$$R(z) = \sum_{r \geqslant 0} R_r z^r = \frac{1 - z + z^2}{(1 - z)^3}.$$

A. Camina, B. Lewis, *An Introduction to Enumeration*,
Springer Undergraduate Mathematics Series,
DOI 10.1007/978-0-85729-600-9_3, © Springer-Verlag London Limited 2011

But we can expand the denominator by the Binomial Theorem and then

$$\sum_{r \geqslant 0} R_r z^r = \left(1 - z + z^2\right)\left(1 - z\right)^{-3} = \left(1 - z + z^2\right) \sum_{r \geqslant 0} \binom{r+2}{2} z^r.$$

In effect this gives us an explicit form for the number of regions – a major break-through. All we need to do is to write out (as far as we please) the power series on the right:

$$(1 - z + z^2)\left(\binom{2}{2} + \binom{3}{2}z + \binom{4}{2}z^2 + \cdots\right)$$

$$= \binom{2}{2} + \binom{3}{2}z + \binom{4}{2}z^2 + \binom{5}{2}z^3 + \cdots$$

$$- \binom{2}{2}z - \binom{3}{2}z^2 - \binom{4}{2}z^3 + \cdots$$

$$+ \binom{2}{2}z^2 + \binom{3}{2}z^3 + \binom{4}{2}z^4 + \cdots$$

$$= 1 + (3 - 1)z + (6 - 3 + 1)z^2 + (10 - 6 + 3)z^3 + \cdots.$$

It is easy to see how the general term is formed from Binomial coefficients:

$$R_r = \binom{r+2}{2} - \binom{r+1}{2} + \binom{r}{2} = \tfrac{1}{2}\left(r^2 + r + 2\right).$$

We often refer to this final part as **"comparing coefficients of z^r on either side"**. This selects the rth term of the enumerative sequence on the left and an explicit expression for it on the right. This is one of the principal reasons why generating functions are so useful.

Example 3.2

We expand as a power series the function given by $(1 + z)/(1 - 3z)^2$.

$$\sum_{r \geqslant 0} a_r z^r = \frac{1 + z}{(1 - 3z)^2} = (1 + z)(1 - 3z)^{-2};$$

then we may expand the final term by the Binomial Theorem which gives

$$= (1 + z) \sum_{r \geqslant 0} (r + 1)3^r z^r.$$

This function generates the sequence $\{a_r\}$ and when we compare coefficients of z^r on either side we have

$$a_r = (r + 1)3^r + r3^{r-1} = (4r + 3)3^{r-1}$$

which is true for $r \geqslant 0$.

These two examples show that a generating function whose denominator is a power of a linear expression can certainly be expanded – in a particularly simple way. Unfortunately, not all generating functions are of this form.

Example 3.3 (Generating functions with a quadratic denominator)

Suppose we are given the generating function (whose denominator is quadratic) for the sequence $\{u_r\}$

$$\sum_{r \geq 0} u_r z^r = \frac{5 - 13z}{1 - 5z + 6z^2}.$$

We notice that the denominator factorizes (over the integers) so that

$$\sum_{r \geq 0} u_r z^r = \frac{5 - 13z}{(1 - 2z)(1 - 3z)}.$$

Once more partial fractions come to our aid

$$\sum_{r \geq 0} u_r z^r = \frac{3}{1 - 2z} + \frac{2}{1 - 3z}.$$

If we expand each term by the Binomial Theorem, we have

$$\sum_{r \geq 0} u_r z^r = 3 \sum_{r \geq 0} 2^r z^r + 2 \sum_{r \geq 0} 3^r z^r$$

and comparing coefficients of z^r gives the explicit form

$$u_r = 3.2^r + 2.3^r.$$

But what if the denominator does not factorize? We now investigate such a generating function – that for the Lucas sequence (Exercise 2.14).

Example 3.4

We start with the generating function of the Lucas sequence $\{L_r\}$ which is

$$L(z) = \sum_{r \geq 0} L_r z^r = \frac{2 - z}{1 - z - z^2}.$$

The denominator cannot be factorized (over the integers) but we can write

$$1 - z - z^2 = (1 - \phi_1 z)(1 - \phi_2 z) \tag{3.1}$$

and comparing coefficients of z^r (for $r = 1, 2$) we must have $\phi_1 + \phi_2 = 1$ and $\phi_1 \phi_2 = -1$. Using these relations we may write the generating function in the form

$$L(z) = \sum_{r \geq 0} L_r z^r = \frac{2 - z}{(1 - \phi_1 z)(1 - \phi_2 z)}.$$

Then using standard partial fraction techniques we find that

$$L(z) = \sum_{r \geq 0} L_r z^r = \frac{1}{1 - \phi_1 z} + \frac{1}{1 - \phi_2 z}.$$

Now we can expand each term on the right

$$\sum_{r \geq 0} L_r z^r = \left(1 + \phi_1 z + \phi_1^2 z^2 + \cdots\right) + \left(1 + \phi_2 z + \phi_2^2 z^2 + \cdots\right) = \sum_{r \geq 0} \left(\phi_1^r + \phi_2^r\right) z^r$$

and so on comparing powers of z^r we have the formula

$$L_r = \phi_1^r + \phi_2^r.$$

We can now make this explicit, by solving the two relations that come from Equation (3.1). These reduce to a quadratic whose roots are $\phi_1 = \frac{1 + \sqrt{5}}{2}$ and $\phi_2 = \frac{1 - \sqrt{5}}{2}$. We have proved the theorem that follows.

Theorem 3.5 (Binet's formula for the Lucas sequence)

Each term of the sequence is given by an expression called Binet's formula for the Lucas numbers:

$$L_r = \phi_1^r + \phi_2^r = \frac{\left(1 + \sqrt{5}\right)^r + \left(1 - \sqrt{5}\right)^r}{2^r}.$$

The Lucas sequence $\{L_r\}$ has a celebrity sibling, the Fibonacci sequence $\{F_r\}$. They share many properties. We derive a corresponding Binet formula for them.

Example 3.6

The generating function for the Fibonacci sequence may be factorized (as in the last example)

$$\sum_{r \geq 0} F_r z^r = \frac{z}{(1 - \phi_1 z)(1 - \phi_2 z)}$$

if we use partial fractions on this we find that

$$\sum_{r\geq 0} F_r z^r = \frac{1}{\phi_1 - \phi_2}\left(\frac{1}{1-\phi_1 z} - \frac{1}{1-\phi_2 z}\right) = \frac{1}{\sqrt{5}}\left(\frac{1}{1-\phi_1 z} - \frac{1}{1-\phi_2 z}\right).$$

We can expand the two terms on the right (note that $\phi_1 - \phi_2 = \sqrt{5}$) and so

$$\sum_{r\geq 0} F_r z^r = \frac{1}{\sqrt{5}}\left(1+\phi_1 z + \phi_1^2 z^2 + \cdots\right) - \frac{1}{\sqrt{5}}\left(1+\phi_2 z + \phi_2^2 z^2 + \cdots\right)$$

$$= \frac{1}{\sqrt{5}}\sum_{r\geq 0}(\phi_1^r - \phi_2^r)z^r$$

and so on comparing powers of z^r we have,

$$F_r = \frac{\phi_1^r - \phi_2^r}{\sqrt{5}}.$$

Written in full, this is the explicit formula which proves the theorem that follows.

Theorem 3.7 (Binet's formula for the Fibonacci sequence)

Each term of the Fibonacci sequence is given by an expression also called Binet's formula, this time for the Fibonacci numbers:

$$F_r = \frac{1}{\sqrt{5}}(\phi_1^r - \phi_2^r) = \frac{\left(1+\sqrt{5}\right)^r - \left(1-\sqrt{5}\right)^r}{2^r\sqrt{5}}.$$

Example 3.8 (Fibonacci numbers and Binomial coefficients)

We can expand the generating function for the Fibonacci numbers without using partial fractions. The generating function for the Fibonacci numbers may be written in this way

$$\sum_{r\geq 0} F_r z^r = \frac{z}{1-z-z^2} = \frac{z}{1-z(1+z)} = z(1-z(1+z))^{-1}$$

and on expansion as a power series by the Binomial Theorem (twice: *first* for a negative power, and *then* as a positive power), this becomes

$$\sum_{r\geq 0} F_r z^r = \sum_{s\geq 0} z^{s+1}(1+z)^s = \sum_{s\geq 0} z^{s+1}\sum_{t\geq 0}\binom{s}{t}z^t = \sum_{s\geq 0}\left(\sum_{t\geq 0}\binom{s}{t}\right)z^{s+t+1}.$$

Comparing powers of z^r on either side, so that

$$r = s+t+1 \Rightarrow s = r-t-1$$

we deduce a surprising result.

Theorem 3.9 (Fibonacci and Binomial coefficients)

The Fibonacci numbers are sums of Binomial coefficients:

$$F_r = \sum_{t \geq 0} \binom{r-t-1}{t} = \binom{r-1}{0} + \binom{r-2}{1} + \binom{r-3}{2} + \cdots + \begin{cases} 1 \text{ if } r \text{ is odd;} \\ \frac{r}{2} \text{ otherwise.} \end{cases}$$

This remarkable result shows that the Fibonacci numbers are the sums of diagonal entries in Pascal's triangle.

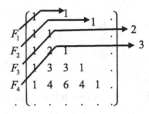

Figure 3.1 Diagonal sums in Pascal's triangle.

3.1.1 Exercises

Exercise 3.1

The sequence $\{v_r\}$ has the generating function

$$V(z) = \frac{1+3z}{1-4z}.$$

Expand the generating function to find an expression for each term of the sequence.

Exercise 3.2

Check that Binet's formulas for the Fibonacci and Lucas sequences do produce integers for the first three terms of each sequence.

Exercise 3.3

The sequence $\{u_r\}$ has the generating function

$$U(z) = \frac{2-3z}{1-3z+z^2}.$$

By factorizing the denominator (not necessarily over the integers) find a Binet style formula for the terms of the sequence. Use it to find the first five terms. What connection does this sequence have with the Lucas sequence?

Exercise 3.4

A sequence $\{v_r\}$ has the generating function

$$V(z) = \frac{1+z}{1-3z+z^2}.$$

Find an explicit form for the terms of the sequence. What is the connection between this sequence and the Fibonacci sequence?

3.2 It's the Denominator that Counts

There is another use of generating functions that we have not yet looked at. A generating function may be used to find a recurrence relation for the sequence it represents.

Example 3.10

Given the generating function for a sequence $\{u_r\}$:

$$\sum_{r \geqslant 0} u_r z^r = \frac{5-13z}{1-5z+6z^2}$$

we take the denominator over to the other side

$$\left(1-5z+6z^2\right) \sum_{r \geqslant 0} u_r z^r = 5-13z.$$

If we now compare coefficients of z^r on either side we find that $u_0 = 5$, $u_1 - 5u_0 = -13$ and then for $r \geqslant 2$ we have the recurrence

$$u_r - 5u_{r-1} + 6u_{r-2} = 0.$$

It follows that the sequence satisfies the recurrence $u_r = 5u_{r-1} - 6u_{r-2}$ with the initial values $u_0 = 5$ and $u_1 = 12$.

But there are some surprises in store.

Example 3.11

Consider the sequence $\{a_r\}$ with the generating function

$$\sum_{r \geqslant 0} a_r z^r = \frac{3z}{1 - 3z + 2z^2}.$$

The denominator factorizes

$$\sum_{r \geqslant 0} a_r z^r = \frac{3z}{(1 - z)(1 - 2z)}$$

and if we take the second factor over to the left-hand side, we have

$$(1 - 2z) \sum_{r \geqslant 0} a_r z^r = \frac{3z}{(1 - z)} = 3z \sum_{r \geqslant 0} z^r$$

and then comparing coefficients of z^r gives

$$a_r - 2a_{r-1} = 3$$

which is the recurrence

$$a_r = 2a_{r-1} + 3.$$

If we had taken the other factor across we would have found another recurrence

$$a_r = a_{r-1} + 3.2^{r-1}.$$

Of course, had we taken the whole denominator across, we would have obtained the recurrence

$$a_r - 3a_{r-1} + 2a_{r-2} = 0 \text{ for } r \geqslant 2.$$

Definition 3.12 (Rational function)

A rational function is of the form $\frac{A(z)}{B(z)}$ where the numerator and denominator are both polynomials.

You may have noticed two intriguing things about the examples already encountered. The first is that the denominator of the generating function (if it is a rational function) has the same "form" as one of the recurrences associated with the corresponding sequence. The second is that a given sequence often has more than one recurrence associated with it. We explore these observations.

Example 3.13

Using these ideas we may find the recurrence satisfied by the sequence $\{F_0, F_5, F_{10}, \cdots\}$ consisting of those Fibonacci numbers whose index is a multiple of 5. The generating function for this sequence is easily found from Binet's formula:

$$
\sum_{r \geq 0} F_{5r} z^r = \sum_{r \geq 0} \frac{1}{\sqrt{5}} \left(\phi_1^{5r} - \phi_2^{5r} \right) z^r = \frac{1}{\sqrt{5}} \left(\sum_{r \geq 0} \phi_1^{5r} z^r - \sum_{r \geq 0} \phi_2^{5r} z^r \right)
$$

$$
= \frac{1}{\sqrt{5}} \left(\sum_{r \geq 0} \left(\phi_1^5 z \right)^r - \sum_{r \geq 0} \left(\phi_2^5 z \right)^r \right).
$$

Each of the sums is an infinite GP (convergent – for suitably restricted values of z) and hence:

$$
\sum_{r \geq 0} F_{5r} z^r = \frac{1}{\sqrt{5}} \left(\frac{1}{1 - \phi_1^5 z} - \frac{1}{1 - \phi_2^5 z} \right) = \frac{1}{\sqrt{5}} \left(\frac{(\phi_1^5 - \phi_2^5) z}{1 - (\phi_1^5 + \phi_2^5) z + \phi_1^5 \phi_2^5 z^2} \right)
$$

$$
= \left(\frac{\frac{1}{\sqrt{5}} (\phi_1^5 - \phi_2^5) z}{1 - (\phi_1^5 + \phi_2^5) z + (\phi_1 \phi_2)^5 z^2} \right) = \frac{F_5 z}{1 - L_5 z + (-1)^5 z^2} = \frac{5z}{1 - 11z - z^2}.
$$

The denominator leads immediately to the recurrence,

$$
F_{5r} = 11 F_{5(r-1)} + F_{5(r-2)} \text{ for } r \geq 2.
$$

This is a particular example of an important general result that we now state – you are asked to prove it in Exercise 3.18.

Theorem 3.14 (Rational generating functions)

If the generating function for a sequence is a rational function, then a recurrence satisfied by the sequence corresponds to the polynomial that forms its denominator.

And now for another example that spawns multiple recurrences.

Example 3.15

We know the generating function for the number of regions created when r lines are drawn in the plane – no two of which are parallel, and no three of which are coincident (Example 2.27). It is

$$
\sum_{r \geq 0} R_r z^r = \frac{1 - z + z^2}{(1 - z)^3}.
$$

We may rearrange the generating function like this

$$(1-z)^3 \sum_{r \geqslant 0} R_r z^r = 1 - z - z^2$$
$$\Rightarrow (1 - 3z + 3z^2 - z^3) \sum_{r \geqslant 0} R_r z^r = 1 - z - z^2.$$

Next we compare coefficients of z^r (for $r \geqslant 3$); we find the recurrence

$$R_r = 3R_{r-1} - 3R_{r-2} + R_{r-3} \text{ for } r \geqslant 3.$$

In this example there is a natural link between the recurrence relation and the denominator as demanded by Theorem 3.14 and illustrated in Figure 3.2.

$$\boxed{(1-z)^3 = 1 - 3z + 3z^2 - z^3} \quad \Longleftrightarrow \quad \boxed{R_r - 3R_{r-1} + 3R_{r-2} - R_{r-3} = 0}$$

Figure 3.2 Denominator and recurrence.

However, a number of other recurrences may also be associated with the sequence through its generating function. We can develop three quite different – but not independent – recurrences from this generating function.
The **first** comes in this way

$$(1-z) \sum_{r \geqslant 0} R_r z^r = \frac{1 - z + z^2}{(1-z)^2} = (1 - z + z^2)(1-z)^{-2}.$$

Expanding the right-hand side and then comparing coefficients of z^r we find that,

$$R_r - R_{r-1} = \binom{r+1}{1} - \binom{r}{1} + \binom{r-1}{1} = r.$$

That, of course, was the recurrence we started with. The **second** comes from

$$(1-z)^2 \sum_{r \geqslant 0} R_r z^r = \frac{1 - z + z^2}{(1-z)} = (1 - z + z^2)(1-z)^{-1}$$
$$\Rightarrow R_r - 2R_{r-1} + R_{r-2} = 1 - 1 + 1 = 1 \text{ which is the recurrence}$$
$$R_r - 2R_{r-1} + R_{r-2} = 1.$$

Finally, there is the recurrence already derived from the whole denominator

$$(1-z)^3 \sum_{r \geqslant 0} R_r z^r = 1 - z - z^2$$
$$\Rightarrow R_r - 3R_{r-1} + 3R_{r-2} - R_{r-3} = 0.$$

Definition 3.16 (Linear homogeneous recurrence)

A recurrence is said to be linear homogeneous if it involves only terms of the sequence with constant coefficients (that is coefficients independent of the index r).

Example 3.17

We may classify – as linear or not – the recurrences of the last example. The first recurrence $R_r - R_{r-1} = r$ is not linear because of the "r" term; neither is the second recurrence $R_r - 2R_{r-1} + R_{r-2} = 1$ because of the "1". The final recurrence

$$R_r - 3R_{r-1} + 3R_{r-2} - R_{r-3} = 0$$

is, however, linear homogeneous since it can be written $R_r = 3R_{r-1} - 3R_{r-2} + R_{r-3}$.

Theorem 3.18

A sequence whose generating function is rational, with a denominator that may be factorized over the integers, obeys the recurrences shown in Figure 3.3.

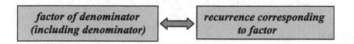

Figure 3.3 Factor of denominator and recurrence.

A final point – we don't need to restrict the factors to those with integer coefficients.

Example 3.19

By factorizing the denominator of the generating function for the Fibonacci sequence in this way

$$\sum_{r \geqslant 0} F_r z^r = \frac{z}{(1 - \phi_1 z)(1 - \phi_2 z)},$$

we get

$$(1 - \phi_1 z) \sum_{r \geqslant 0} F_r z^r = \frac{z}{(1 - \phi_2 z)} = z \sum_{r \geqslant 0} \phi_2^r z^r.$$

By comparing coefficients of z^r we derive the "recurrence"

$$F_r = \phi_1 F_{r-1} + \phi_2^{r-1}.$$

There is another way to associate more recurrences with a sequence. Suppose that the polynomial making up the denominator of the rational generating function for the sequence $\{u_r\}$ is $B(z)$. Then we have

$$B(z) \sum_{r \geq 0} u_r z^r = 0 \text{ for } r \geq m+1,$$

where m is the degree of the numerator polynomial $A(z)$, and so

$$f(z)B(z) \sum_{r \geq 0} u_r z^r = 0 \text{ for } r \geq m+1$$

where f is an arbitrary polynomial. For instance in Example 3.13 we have

$$\left(1 - 11z - z^2\right) \sum_{r \geq 0} F_{5r} z^r = 5z$$

$$\Rightarrow (1 + 3z)\left(1 - 11z - z^2\right) \sum_{r \geq 0} F_{5r} z^r = 5z$$

$$\left(1 - 8z - 34z^2 - 3z^3\right) \sum_{r \geq 0} F_{5r} z^r = 5z.$$

So it follows that

$$F_{5r} = 8F_{5(r-1)} + 34F_{5(r-2)} + 3F_{5(r-3)} \text{ for } r \geq 3.$$

The function $f(z) = (1 + 3z)$ used here is quite arbitrary. We can construct any number of recurrences in this way. It is not true, however, that two different recurrences associated with a given sequence are always generated in this way. Particular examples are the recurrences for the derangement sequence that we met in Exercise 2.8.

3.2.1 Exercises

Exercise 3.5

This question concerns the sequence $\{L_{2r}\}$ which consists of the Lucas numbers whose index is a multiple of 2. Write down an explicit (Binet) form for L_{2r} and use this to find a generating function for the sequence. Hence, derive a recurrence relation that the sequence satisfies. Repeat for the sequence $\{L_{3r}\}$.

Exercise 3.6

The sequence $\{u_r\}$ satisfies the recurrence relation $u_r = 2u_{r-1} + 2^r + 3^r$ with the initial term $u_0 = 1$. Find a generating function for the sequence and hence a linear homogeneous recurrence relation that it satisfies.

3.3 Solving Linear Homogeneous Recurrences of Degree 2

We will frequently encounter linear homogeneous recurrence relations for a sequence $\{u_r\}$ of the form

$$u_r = au_{r-1} + bu_{r-2},$$

in which a and b are independent of r. In this section we develop general results that enable us to write down the terms of this sequence in an explicit form. The recurrence is said to be of degree 2.

Definition 3.20 (Degree of a homogeneous linear recurrence)

The linear homogeneous recurrence is said to be of degree k when $\alpha u_r + \beta u_{r-1} + \cdots \zeta u_{r-k} = 0$ (in which α and $\zeta \neq 0$).

Lemma 3.21

The generating function of the sequence $\{u_r\}$ which satisfies the linear homogeneous recurrence of degree 2, $u_r = au_{r-1} + bu_{r-2}$ has the generating function

$$U(z) = \frac{u_0 + (u_1 - au_0)z}{1 - az - bz^2}.$$

Proof

Using the three-step recipe (Algorithm 2.23) for this recurrence leads to a generating function $U(z)$ that satisfies,

$$U(z) - u_0 - u_1 z = az(U(z) - u_0) + bz^2 U(z)$$

which may be solved to give the result stated.   □

Now we bring on the machinery of partial fractions – just as we did with the Lucas numbers in Example 3.4. Suppose that we write the denominator of the generating function $U(z)$ in the form

$$1 - az - bz^2 \equiv (1 - \theta_1 z)(1 - \theta_1 z)$$

then we must have $\theta_1 + \theta_2 = a$ and $\theta_1 \theta_2 = -b$. These latter equations combine to give

$$\theta^2 - a\theta - b = 0$$

and this quadratic has the explicit solutions

$$\theta_1 = \frac{a + \sqrt{a^2 + 4b}}{2} \quad \text{and} \quad \theta_2 = \frac{a - \sqrt{a^2 + 4b}}{2}.$$

Definition 3.22 (Characteristic equation)

The equation $\theta^2 - a\theta - b = 0$ is called the characteristic equation of the sequence $\{u_r\}$.

This equation has distinct roots (real or complex) whenever $\sqrt{a^2 + 4b} \neq 0$ and then if we write

$$U(z) = \sum_{r \geqslant 0} u_r z^r = \frac{A}{(1 - \theta_1 z)} + \frac{B}{(1 - \theta_2 z)},$$

we can easily determine the values of A and B by using suitable values of z. We find that

$$U(z) = \sum_{r \geqslant 0} u_r z^r = \frac{1}{\theta_1 - \theta_2} \left(\frac{u_0 \theta_1 + u_1 - a u_0}{1 - \theta_1 z} - \frac{u_0 \theta_2 + u_1 - a u_0}{1 - \theta_2 z} \right).$$

We may derive an explicit expression (Binet-type) for u_r by expanding the denominators of the two expressions inside the bracket, and then comparing coefficients of z^r on either side. The resulting expression involves powers of θ_1 and θ_2. This, however, is not as useful as it may appear. What is of practical use is another much simplified explicit form,

$$u_r = C\theta_1^r + D\theta_2^r,$$

in which C and D are constants determined by the initial values of the sequence and the values of θ_1 and θ_2.

Theorem 3.23 (Solution of linear homogeneous recurrence of degree 2)

If the recurrence satisfied by the sequence $\{u_r\}$ is $u_r = a u_{r-1} + b u_{r-2}$, where a and b are independent of r and $a^2 + 4b \neq 0$, then:

(i) the explicit solution of the recurrence is $u_r = C\theta_1^r + D\theta_2^r$ where θ_1 and θ_2 are the (distinct) roots of the characteristic equation;

(ii) the values of the constants C and D are determined from the initial terms of the sequence.

Example 3.24

We use this result to solve the recurrence of Example 2.17 which is $s_{r+1} = 3s_r + 3s_{r-1}$ with the initial values $s_0 = 1$ and $s_1 = 4$. The characteristic equation is $\theta^2 - 3\theta - 3 = 0$, which has the solutions

$$\theta_1 = \frac{3+\sqrt{21}}{2} \text{ and } \theta_2 = \frac{3-\sqrt{21}}{2}.$$

So by the previous theorem

$$s_r = C\left(\frac{3+\sqrt{21}}{2}\right)^r + D\left(\frac{3-\sqrt{21}}{2}\right)^r.$$

Now we make use of the given initial values, so that

$$1 = C+D \text{ and } 4 = C\left(\frac{3+\sqrt{21}}{2}\right) + D\left(\frac{3-\sqrt{21}}{2}\right).$$

Solving these we arrive at

$$s_r = \left(\frac{5+\sqrt{21}}{2\sqrt{21}}\right)\left(\frac{3+\sqrt{21}}{2}\right)^r - \left(\frac{5-\sqrt{21}}{2\sqrt{21}}\right)\left(\frac{3-\sqrt{21}}{2}\right)^r.$$

Note:

(i) A **common error** in finding the values of C and D is to work from a single value,

$$4 = C\left(\frac{3+\sqrt{21}}{2}\right) + D\left(\frac{3-\sqrt{21}}{2}\right)$$

and then to equate rational and irrational parts on both sides to give

$$\frac{3}{2}C + \frac{3}{2}D = 4 \text{ and } \frac{\sqrt{21}}{2}C - \frac{\sqrt{21}}{2}D = 0.$$

These simultaneous equations give the solution $C = D = \frac{4}{3}$, which is incorrect.

(ii) **Do not** make the assumption that C and D are necessarily rational or even real!

(iii) The formula for the terms of this sequence $\{s_r\}$ must give integers – despite its appearance.

Of course it can happen that the roots of the characteristic equation are not distinct – but that is easily dealt with.

Example 3.25

We solve the recurrence relation $u_r = 2u_{r-1} - u_{r-2}$ with the initial terms $u_0 = 0$ and $u_1 = 1$. The characteristic equation does not have distinct roots since it is $\theta^2 - 2\theta +$

$1 = 0$, which factorizes as $(\theta - 1)^2 = 0$. We need in this case to return to generating functions; Algorithm 2.23 now gives

$$\sum_{r \geq 2} u_r z^r = 2 \sum_{r \geq 2} u_{r-1} z^r - \sum_{r \geq 2} u_{r-2} z^r$$

so that in the usual notation

$$U(z) - 0 - 1z = 2z(U(z) - 0) - z^2 U(z)$$

and hence

$$U(z) = \frac{z}{1 - 2z + z^2} = \frac{z}{(1-z)^2}.$$

But this is simplicity itself to expand, and then

$$U(z) = \sum_{r \geq 0} u_r z^r = z \sum_{r \geq 0} (r+1) z^r$$

so that $u_r = r$.

3.3.1 Exercises

Exercise 3.7

Solve the recurrence relation $u_r = 2u_{r-1} - 2u_{r-2}$ with the initial values $u_0 = 1$ and $u_1 = 3$.

Exercise 3.8

Show that the recurrence of Example 2.9, $n_{r+1} = 3.10^r + 4n_r$, may be written in the linear homogeneous form $n_{r+1} = 14n_r - 40n_{r-1}$. Use either of these to find an explicit form for n_r.

Exercise 3.9

A Fibonacci-type sequence is defined by the recurrence

$$G_r = G_{r-1} + G_{r-2}$$

with initial values $G_0 = 3$ and $G_1 = 0$. Find an explicit form for its terms using Fibonacci numbers.

Exercise 3.10

Two integer sequences $\{a_r\}$ and $\{b_r\}$ are defined by

$$a_r + b_r \sqrt{3} = (1 + \sqrt{3})^r.$$

Show that $a_{r+1} = a_r + 3b_r$ and $b_{r+1} = a_r + b_r$ and hence:

(i) find the single recurrence (with appropriate initial values) that each sequence satisfies;

(ii) find an explicit form for the terms of each sequence.

Exercise 3.11

In Exercise 2.20 you found a recurrence relation for the number of tilings: $f_r = 2f_{r-1} + 3f_{r-2}$. Find an explicit form for the number of tilings.

3.4 Miscellaneous Exercises

Exercise 3.12

The sequence $\{a_r\}$ is made up of terms in which $a_r = 3a_{r-1} + 2^r$ for $r \geqslant 1$, and initial term $a_0 = 1$. Convert this into a generating function and hence determine an explicit form for each of its terms.

Exercise 3.13

In a permutation π of the set $\{1, 2, 3, \cdots r\}$ the "distance" an element i moves is given by the expression $|\pi(i) - i|$. If $\{u_r\}$ counts the number of permutations that move no more than one unit, find an explicit form for the terms of this sequence.

Exercise 3.14

Let w_r be the number of "words" (alphabetic strings) of length r formed from the letters $\{e, a, t\}$ in which e is always followed by a. Construct the recurrence relation satisfied by the sequence $\{w_r\}$ and hence find an explicit form for its terms.

Exercise 3.15

Suppose that there are w_r "words" of length r made up from the alphabet $\{v, e, r, y\}$, in which there are no consecutive repetition of letters. Find a recurrence for the sequence $\{w_r\}$ and hence find an explicit form for its terms.

Exercise 3.16

Given that $\phi_1 + \phi_2 = 1$, and $\phi_1 \phi_2 = -1$ where ϕ_1 and ϕ_2 are as defined in Example 3.4, show that $\phi_1^2 = \phi_1 + 1$ and find the corresponding result for ϕ_2. Hence prove that $\phi_1^r = F_r \phi_1 + F_{r-1}$.

Exercise 3.17

The recurrence

$$F_r(w) = wF_{r-1}(w) + F_{r-2}(w)$$

with the initial terms $F_0(w) = 1$ and $F_1(w) = w$ defines a sequence of **polynomials** $\{F_r(w)\}$. Show that:

(i) $F_2(w) = w^2 + 1$ and $F_3 = w^3 + 2w$;

(ii) the characteristic equation of the polynomial sequence is

$$\theta^2 - w\theta - 1 = 0;$$

(iii) the polynomials are given by

$$F_r(w) = \frac{\left(w + \sqrt{w^2 + 4}\right)^{r+1} - \left(w - \sqrt{w^2 + 4}\right)^{r+1}}{2^{r+1}\sqrt{w^2 + 4}};$$

(iv) $F_{r+1} = F_r(1)$.

Exercise 3.18

Prove Theorem 3.14.

Exercise 3.19

Let n_r denote the number of non-negative integer solutions of the equation $x + 2y + 3z = r$. By using partial fractions to rewrite the expression,

$$\frac{1 - z^6}{(1 - z)(1 - z^2)(1 - z^3)},$$

show that n_r obeys the surprising recurrence

$$n_r - n_{r-6} = r \ (r \geqslant 7).$$

4

Permutation Groups

4.1 Introduction to Groups

It might seem odd that we want to discuss group theory in a book on enumeration but it will turn out to be an important tool. If we are counting the number of arrangements of a set of objects, we might wish to count two arrangements which are symmetrical as being the same. A simple example is that of a bracelet made up of different coloured beads. It then makes sense to count two arrangements of the beads as the same if one can be rotated into the other.

We will assume a basic knowledge of the definition of groups as well as an ability to use them in a simple way. This can all be found in one or two books as part of this series. So we assume that readers are familiar with the following definitions.

Definition 4.1 (Subgroup)

Let G be a group and let H be a subset of G. If H is a group with respect to the same law of composition then H is called a subgroup of G.

Definition 4.2 (Order of group)

Let G be a group. If the number of elements in G, denoted by $|G|$, is finite, the group is said to be a **finite group**. Otherwise the group is said to be infinite. If G is a finite group then the order of G is the size of the set G. If $g \in G$, then g is said to have **finite order** if the subgroup generated by g is finite.

A. Camina, B. Lewis, *An Introduction to Enumeration*,
Springer Undergraduate Mathematics Series,
DOI 10.1007/978-0-85729-600-9_4, © Springer-Verlag London Limited 2011

Definition 4.3 (Normal subgroup)

Let G be a group and let H be a subgroup of G. If $gH = Hg$ for all $g \in G$, then H is called a normal subgroup of G.

Definition 4.4 (Homomorphism)

Let G and H be groups. A homomorphism from G to H is a mapping $\phi : H \to H$ so that $\phi(xy) = \phi(x)\phi(y)$ for all pairs $x, f \in G$. The set $\{x \in G : \phi(x) = 1\}$ is called the **kernel** of ϕ.

If ϕ is one–one and onto then ϕ is called an **isomorphism**.

An important observation is that the kernel is a normal subgroup of G: see Exercise 4.3.

Definition 4.5 (Cosets)

Let G be a group and let H be a subgroup of G. Let $x \in G$ and consider the set $Hx = \{hx : h \in H\}$. This is called a **right coset** of H in G.

Similarly we define a **left coset** as $xH = \{xh : h \in H\}$. The term coset means either a left coset or a right coset.

Theorem 4.6

Let G be a group and H a subgroup of G. Then the cosets of H partition G into sets of equal cardinality. Thus if G is finite, $|H|$ divides $|G|$.

The number of cosets is called the **index** of H in G. This is usually denoted by $[G : H]$.

This result is sometimes known as **Lagrange's Theorem**, though there is no evidence that he either knew or proved it. We will prove this in Proposition 4.49.

4.1.1 Exercises

Exercise 4.1

Let G be a group. Show that the intersection of subgroups is a subgroup. Hence show that for any subset X there is a unique smallest subgroup containing X. This is called the subgroup generated by X and is written $\langle X \rangle$.

Exercise 4.2

Let G be a group. Show that $g \in G$ has order $n \in \mathbb{N}$, if and only if n is the least positive integer such that $g^n = 1$.

Exercise 4.3

Let G and H be groups and let $\phi : G \to H$ be a homomorphism between them.

(i) Show that the kernel of ϕ is a normal subgroup.

(ii) Show that if $g \in G$ has order n then order of $\phi(g)$ divides n.

Exercise 4.4

Let G be a group and assume the order of G is a prime number. Show that G has only two subgroups.

Exercise 4.5

Let p be a prime number. Prove that any two groups of order p are isomorphic.

4.2 The Symmetric Group

In this section we define the symmetric group and look at some of its properties. We begin by noting the following result.

Lemma 4.7

The composition of functions is associative.

Proof

Let V, X, Y and Z be sets and let $f : V \to X$, $g : X \to Y$ and $h : Y \to Z$ all be functions. Choose $y \in Y$ then

$$((hg)f)(y) = (hg)(f(y)) = (h(g(f(y))) = ((h)(gf)(y) = ((h(gf)(y)).$$

All this says is that if we map y by h and then by fg it is the same as mapping y by gh and then by f, which is just the same as mapping y by f, then by g and finally by h. $\qquad\square$

Note: writing composition of functions as $\circ$ this says that $h \circ (g \circ h) = (h \circ g) \circ (f)$.

Proposition 4.8

Let X be a set. The set of permutations of X with multiplication defined by composition of functions is a group.

The proof is set as Exercise 4.11. This property gives rise to the following definition.

Definition 4.9 (Symmetric group)

Let X be a set. The group of all permutations on X is called the symmetric group on X. This is denoted by $\mathrm{Sym}(X)$. The **degree** of the group is $|X|$. When $X = \{1, 2, 3, \ldots, n\}$ for some $n \in \mathbb{N}$ we write $\mathrm{Sym}(n)$ to denote the symmetric group on a set with n elements.

Example 4.10

Obviously if $|X| = 1$ the group has only one element. A group with only one element is called the **trivial** group.

Example 4.11

What happens if $|X| = 2$? Assume $X = \{1, 2\}$. The group has only two elements. One element is just the identity map; the other, say g, is given by $g(1) = 2$ and $g(2) = 1$. All very straightforward.

We usually think of X as the set $\{1, 2, 3, \ldots n\}$ for some $n \in \mathbb{N}$. There are two main ways to write a permutation, π. One is to write the elements along a line and the images underneath; this is called the bijective form (see page 4)

$$\begin{pmatrix} 1 & 2 & \cdots & n \\ a_1 & a_2 & \cdots & a_n \end{pmatrix}$$

so that $\pi(1) = a_1, \pi(2) = a_2, \cdots, \pi(n) = a_n$.
So

$$\begin{pmatrix} 1 & 2 & 3 & 4 \\ 2 & 4 & 1 & 3 \end{pmatrix}$$

would map $1 \to 2$, $2 \to 4$, $3 \to 1$ and $4 \to 3$. This is quite a clumsy notation.

We will use cycle notation for permutations (see page 4). Let π be a permutation. An example is the permutation above, which we could write as (1243) where each element is mapped to the one on the right, but think of them being written in a circle so that the last goes to the first. Clearly $(1243) = (2431) = (4312) = (3124)$. This is a very compact way to write permutations.

Permutations are defined as functions, so the composition of the permutations π and σ is $\pi \circ \sigma$.

Example 4.12

We evaluate composition of permutations (written in cycle form) using the bijective form as an intermediate step:

(i) the composition $(123)(12)$ maps the sequence $\{123\}$ first to $\{213\}$ and then to $\{321\}$. So the product, in cycle notation, is $(13)(2) = (13)$

(ii) the reverse composition $(12)(123)$ maps the sequence $\{123\}$ to $\{231\}$ and then to $\{132\}$. So the product, in cycle notation, is $(1)(23) = (23)$.

Note: composition of permutations (and functions) is **not** commutative, as this example shows.

Example 4.13

We will write the permutation below in cycle form

$$\pi = \begin{pmatrix} 1 & 2 & 3 & 4 & 5 \\ 2 & 3 & 1 & 5 & 4 \end{pmatrix}.$$

If we start with 1 we may write this as $(123)(45)$. This is read as $1 \to 2 \to 3 \to 1$ and $4 \to 5 \to 4$. We could start with 3 and write this as $(312)(54)$. It usually makes sense to start with the smallest number and work up. Remember that we are writing functions on the left, so we start with the function on the right.

Example 4.14

Let

$$\pi = \begin{pmatrix} 1 & 2 & 3 & 4 & 5 & 6 \\ 2 & 3 & 6 & 4 & 1 & 5 \end{pmatrix}.$$

We write this as (12365) not as $(12365)(4)$.

So if a point is fixed we call it a **fixed point** and we omit it in the cycle notation. It is normally clear, except when we want to describe the identity element or are not sure where the permutation lives.

Example 4.15

Consider this permutation $(123)(45)$. It could be in Sym(5) or Sym(7) where it leaves 6 and 7 fixed. Usually the context makes it clear.

The cycle notation gives us a lot of information about the permutation that it represents. The next lemma shows that we can always write permutations as a product of disjoint cycles; see page 4.

Lemma 4.16

Let $g \in \mathrm{Sym}(n)$. Then g can be written as a product of disjoint cycles, say

$$(a_1 a_2 \ldots a_{r_1})(a_{r_1+1} a_{r_1+2} \ldots a_{r_1+r_2}) \cdots (a_{r_1+\cdots+r_{c-1}} \ldots a_{r_1+\cdots+r_c});$$

Here the sets of entries in each cycle are disjoint and the cycles have lengths r_1, r_2, $\ldots, r_c$. Further the order of g is $\mathrm{lcm}(r_1, r_2, \ldots, r_c)$.

Proof

Choose an element, i, not fixed by g. If there isn't one, then g is the identity so we are done. Begin by writing the cycle $(i\, g(i)\, g^2(i) \ldots)$. We know that at some stage an element will be repeated, as there are only finitely many elements. Let s be the power of g which gives the first repeat. So $g^s(i) = g^t(i)$ for some $0 \le t < s$.

Note that $g^{s-t}(i) = g^0(i) = i$, since g^0 is the identity. Since s was chosen to be the first repeat $t = 0$ and $g^s(i) = i$. Further the points $i, g(i), g^2(i), \ldots, g^{s-1}(i)$ are all distinct. So $(i\, g(i) \ldots g^{s-1}(i))$ is a cycle.

Let $Y = \{1, 2, \ldots, n\} \setminus \{(i, g(i), \ldots, g^{s-1}(i))\}$. If there are no elements in Y moved by g we are done. If not choose $j \in Y$ so that $g(j) \ne j$. Repeat the process where we have u so that $(j, g(j), \ldots, g^{u-1}(j))$.

We now have to show that these two cycles are disjoint. So assume false and then we get $g^a(i) = g^b(j)$ for some a, b, but then $j = g^{a-b}(i)$ so j would be part of the first cycle, which is false.

Let $\pi = (a_1 a_2 \ldots a_r)$ be a cycle length r. Then $\pi^r(a_i) = a_i, 1 \le i \le r$. So π has order r. The final part of the lemma follows immediately.   □

We say that a cycle of length r is an r-**cycle**. A cycle of length 2 is called a **transposition**. The cycle structure of a permutation gives considerable information about a permutation. The next lemma counts how many permutations there are of each cycle structure, but first we count how many permutations there are. We know from Theorem 1.7 that $|\mathrm{Sym}(n)| = n!$.

Example 4.17

Let us count the number of permutations of each cycle type in the symmetric group of degree 4. We denote the cycle type of an element by a sequence i^{n_i} where the element has n_i cycles of length i. The element (12) has cycle type $1^2, 2^1$ as it has two cycles of length 1, namely (3) and (4) and one of length 2, (12).

Cycle type	Elements	Number
1^4	$(1)(2)(3)(4)$	1
$1^2 2^1$	$(12), (13), (14), (23),$ $(24), (34)$	6
2^2	$(12)(34), (13)(24),$ $(14)(32)$	3
$1^1 3^1$	$(123), (132), (134),$ $(143), (124), (142),$ $(234), (243)$	8

Table 4.1 Elements of $\mathrm{Sym}(4)$.

Another way to calculate the entries in this table is to think, for example, of how many permutations there are with two cycles of length 2. So we need to choose elements to replace the dots in $(\bullet\bullet)(\bullet\bullet)$. We can see that there are 4! ways to do this, but many will give the same permutation, for example, $(12)(34) = (12)(43) = (21)(34) = (21)(43)$. Further we can also reverse the order of the product. So there are 8 ways to write this element so we end up with $24/8 = 3$ such elements.

We now have a general result which shows how to count this number.

Theorem 4.18

The number of elements of $\mathrm{Sym}(n)$ which have n_i cycles of length i is

$$\frac{n!}{\prod n_i! \, i^{n_i}}.$$

Proof

Imagine that we have laid out in a row the cycles which have to be filled in to create the permutation with n_i cycles of length i. There are $n!$ ways we can fill in the cycles. However, many of them will give the same permutation. But we can count the number of times this happens. Each cycle of length i can be written in i ways by just cycling the order and there are n_i cycles. This gives the factor i^{n_i}. But there are n_i such cycles and they can be permuted in $n_i!$ factorial ways and the result follows. □

Example 4.19

Let us check the answers in the example above. The second line would give $4!/2!1^21!2^1$ which is $24/4 = 6$. The third line is $24/2!2^2 = 3$.

Another important property of permutations is a number called its **sign**, which as we will see counts the number of transpositions into which it can be decomposed.

Lemma 4.20

Let $g \in \mathrm{Sym}(n)$. Then

$$\prod_{i<j}(g(i) - g(j)) = \varepsilon_g \prod_{i<j}(i - j),$$

where $\varepsilon_g = \pm 1$.

Proof

Since all the possible differences occur exactly once in each product we see that $\varepsilon_g = \pm 1$.   □

Definition 4.21 (Sign of permutation)

The coefficient from the previous lemma, ε_g, is called the sign of g. If $\varepsilon_g = 1$ the g is called an **even permutation** and if $\varepsilon_g = -1$ the g is called an **odd permutation**.

Proposition 4.22

Let $g, h \in \mathrm{Sym}(n)$. Then $\varepsilon_{gh} = \varepsilon_g \varepsilon_h$. The set of even permutations form a normal subgroup of index 2 called the **Alternating Group**.

Proof

We begin by defining a_g for $g \in \mathrm{Sym}(n)$ to be equal to the number of pairs $(i,j) \in X \times X$, $i \neq j$ so that $g(i) - g(j)$ has a different sign from $i - j$. Then $\varepsilon_g = (-1)^{a_g}$. Note that this number is not dependent on the ordering of the set X. So for any pairs $g, h \in G$

$$\varepsilon_{gh} \prod_{i<j}(i - j) = \prod_{i<j}(gh(i) - gh(j))$$

since the ordering is irrelevant, this can be rewritten as

$$\varepsilon_g \prod_{i<j}(h(i) - h(j)) = \varepsilon_g \varepsilon_h \prod_{i<j}(i - j).$$

Now we see that the map $g \to \varepsilon_g$ is a homomorphism from G into the multiplicative group $\{+1, -1\}$. The kernel is exactly the set of even permutations, thus proving the last part of the proposition. $\qquad\square$

Example 4.23

Consider $\mathrm{Sym}(n)$ and a transposition $g = (ij)$. How many pairs, as in the previous proposition, will have their sign changed by the action of g? The answer is the pairs of the form (i, r) and (s, j) where $i < r \leq j$ and $i \leq s < j$. There are $2(j - i) - 1$ such changes, so all transpositions are odd, somewhat oddly.

Definition 4.24 (Cyclic permutation)

A permutation which consists of one cycle of length r and cycles of length 1, is called a cyclic permutation.

Lemma 4.25

A cyclic permutation can be written as as the product of $r - 1$ transpositions. Hence a cyclic permutation of length r is an odd permutation if r is even and is an even permutation if r is odd, where a cyclic permutation is one that consists of one cycle.

Proof

Let the cycle be $(a_1 a_2 \ldots a_r)$. Note that

$$(a_1 a_2 \ldots a_r) = (a_2 a_3) \cdots (a_{r-1} a_r)(a_1 a_r).$$

The result now follows from the previous example and proposition. $\qquad\square$

Since any permutation is a product of cycles then any permutation can be written as a product of transpositions. However, there are many different ways that this can be done. The number always has the same parity because this determines the sign which is an invariant of the permutation.

Lemma 4.26

The alternating group of degree n has order $n!/2$.

Proof

This follows immediately from Theorem 1.7 and Proposition 4.22. $\qquad\square$

Lemma 4.27

Any even permutation can be written as a product of three cycles.

Proof

We know that any permutation can be written as a product of transpositions. If the permutation is even, it can be written as a product of an even number of transpositions. We can group them in pairs and so we need to consider the product of two transpositions, say $(ab)(cd)$. If $\{a,b\} = \{c,d\}$ the product is the identity. If there are only three distinct symbols, we can assume that the product is $(ab)(ac)$. But $(ab)(ac) = (acb)$. If they are all distinct, we get $(ab)(cd) = (acb)(acd)$. So the proof is complete.   □

Definition 4.28 (A permutation group)

Let n be a natural number. A permutation group of degree n is a subgroup of $\mathrm{Sym}(n)$.

A given abstract group can be a permutation group in many different ways, even of the same degree.

Example 4.29

Consider $G = \mathrm{Sym}(4)$. Then we can define two subgroups of G,

 (i) $A = \{1, (12)(34), (13)(24), (14)(23)\}$;

(ii) $B = \{1, (12), (12)(34), (34)\}$.

Both groups are abelian of order 4 and all non-trivial elements have order 2. So the two subgroups are isomorphic as abstract groups. However, they differ as permutation groups: A has an element which maps 1 to 3 but B does not.

4.2.1 Exercises

Exercise 4.6

Show that $\mathrm{Sym}(4)$ has no elements of order 5.

Exercise 4.7

Show that the $\mathrm{Sym}(n)$ has no elements of order p for any prime $p > n$.

Exercise 4.8

For the group $\mathrm{Sym}(5)$ enumerate the different cycle types; calculate the orders of such elements; count how many there are of each type.

Exercise 4.9

Show that the maximal order of an element in $\text{Sym}(10)$ is 30.

Exercise 4.10

Find the generating function for the number of transpositions in the symmetric group.

4.3 Group Actions

Definition 4.30 (Group action)

Let G be a group and X be a set. Then we say that G **acts** on X, if for each $g \in G$ there is a mapping $\phi(g) : X \to X$ so that

(i) $\phi(g)\phi(h)(x) = \phi(gh)(x), \forall x \in X$ and $\forall g, h \in G$ and

(ii) $\phi(1)(x) = x, \forall x \in X$.

We see with this definition that $\phi(G)$ is a permutation group of degree $|X|$.

Example 4.31

If we take X as the set $X = \{1, 2, 3, \ldots, n\}$ and the group G as the symmetric group on X. Then ϕ may be the identity. In fact this works for any permutation group.

Example 4.32 (Regular Action)

In this we make a group act on itself. So both the group and the set are G. We have to define an action ϕ. So we define $\phi(g)(x) = gx$. We also have to check the rules for an action.

$$\phi(g)\phi(h)(x) = \phi(g)(hx) = g(hx) = (gh)(x) = \phi(gh)(x).$$

Also $\phi(1)(x) = 1x = x$.

Example 4.33 (Conjugate Action)

Again we make a group act on itself. So both the group and the set are G. This time however, we define $\phi(g)(x) = gxg^{-1}$. The proof follows the proof in the example above.

$\phi(gh)(x) = ghxh^{-1}g^{-1} = g(hxh^{-1})g^{-1}) = (g\phi(h)(x)g^{-1} = \phi(g)(\phi(h))(x) = \phi(gh)(x).$
Also $\phi(1)(x) = 1x = x.$

Definition 4.34 (Conjugate)

Let G be a group and let x and y be two elements of G. Then x and y are said to be conjugate if there exists $g \in G$ so that $gxg^{-1} = y$.

Note: In many books the definition would be $g^{-1}xg = y$. As we write functions on the right we choose the form $gxg^{-1} = y$. However, they are the same relationship as $(g^{-1})^{-1}xg^{-1} = y$.

Example 4.35

In this we make a group act on subsets of itself, so we define $X = \{Y : Y \subseteq X\}$. Then $\phi(g)(Y) = gY$.

Example 4.36

Consider $G = \text{Sym}(4)$. Then we can define two subgroups of G.

(i) $A = \{1, (12)(34), (13)(24), (14)(23)\}$ and

(ii) $B = \{1, (12), (12)(34), (34)\}$.

This is taken from Example 4.29. Consider polynomials in four indeterminates X_1, X_2, X_3 and X_4 (see Section 2.1) and let A and B act on them as follows:

$$gP(X_1, X_2, X_3, X_4) = P(X_{g(1)}, X_{g(2)}, X_{g(3)}, X_{g(4)})$$

whenever g is in G.

Let $P = X_1X_3 + X_2X_4$ and see what happens as we act on this polynomial. For all $a \in A$ we have $a(P) = P$. If we consider

$$(12)P = X_2X_3 + X_1X_4$$

we see that $(12)P \neq P$. This is another illustration of the difference of the two permutation properties of A and B.

Example 4.37

Let $X_1, X_2, \ldots, X_n$ be n indeterminates (see Section 2.1) and let G be a permutation group of degree n. Let F be a field and let $X = F[X_1, X_2, \ldots, X_n]$. We let G act on X by permuting the suffices, that is if $P(X_1, X_2, \ldots, X_n)$ is a polynomial and $g \in G$ then gP is defined to be the polynomial $P(X_{g(1)}, X_{g(2)}, \ldots, X_{g(n)})$.

This gives an alternative way of defining the sign of a permutation. Let the polynomial P be $\prod(X_i - X_j)$. Then the sign is given by $gP = \varepsilon_g P$, since $g(P) = \pm P$. As G acts on X we get the multiplicative property, see Proposition 4.22.

An important class of examples, which we will be particularly interested in, come from geometry, when we examine the groups of symmetries which fix a particular shape or configuration. These groups are isometries of Euclidean space, that is those mappings which fix distances. Our examples will come from two- and three-dimensional Euclidean space.

Example 4.38

Let G be the group of symmetries of the regular n-gon. First suppose that X is the set of vertices of the n-gon. Then G has a group action on X. G has order $2n$. Also G has a group action on the edges.

Definition 4.39 (Cyclic groups)

A group G with the property that G has an element G so that every element in G is a power of g is called a cyclic.

We now give an example.

Example 4.40

Let C be the group of rotational symmetries, of the regular n-gon. Now the symmetry given by the reflection is not in C. Let ρ be the rotation through $2\pi/n$ radians. Then every element of C is just a power of ρ.

Cyclic groups are particularly easy to study because of the generating element. Let G be cyclic of order n and let g generate G. So the order of g is n. We know from Lagrange's Theorem 4.6 that, if H is a subgroup of G, then $|H|$ divides n. But in this case the converse holds. If $d|n$ then $g^{n/d}$ has order d.

Example 4.41

Let T be a regular tetrahedron and let G be the group of symmetries of T. T has four vertices, six edges and four faces. Thus G can be represented as a permutation group of degree 4 and of degree 6 (see Figure 4.1).

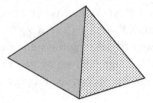

Figure 4.1 Tetrahedron.

Definition 4.42 (Stabilizer and fixed points)

Let G act on a set X. Then:

(i) for $x \in X$ then the stabilizer of x in G is $G_x = \{g \in G : g(x) = x\}$;

(ii) if $g \in G$ the set of fixed points is defined by $\mathrm{Fix}(g) = \{x \in X : g(x) = x\}$.

Lemma 4.43

Let G act on a set X. Then G_x is a subgroup of G for all $x \in G$.

Proof

Note that $1(x) = x$, so $1 \in G_x$. Also, if $g \in G_x$, then $x = 1(x) = g^{-1}g(x) = x$. So $x = g(x) = g^{-1}(x)$ and $g^{-1} \in G_x$. Finally, if $h \in G_x$, then $gh(x) = g(x) = x$. $\qquad\square$

4.3.1 Exercises

Exercise 4.11

Let X be a set. The set of permutations of X with multiplication defined by composition of functions is a group. (This is Proposition 4.8.)

Exercise 4.12

Let G be a finite group. Let X be any subset of G. Show that the set $\{g \in G : gX = X\}$ is a subgroup of G.

Exercise 4.13

Let C be a cyclic group of order n. Let $c \in C$ generate C. Prove the assertions in Example 4.39 that c has order n and that if $d \mid n$ then $c^{n/d}$ has order d.

Exercise 4.14

Show that the symmetry group of the tetrahedron is isomorphic to Sym(4).

Exercise 4.15

Show that in the symmetric group of degree n, conjugacy is determined by the cycle structure, that is, if $g, h \in \text{Sym}(n)$, then g and h are conjugate, if and only if g and h have the some cycle structure.

Exercise 4.16

Consider the action of $\text{Sym}(4)$ on the indeterminates X_1, X_2, X_3, X_4. Find the subgroup of $\text{Sym}(4)$ which leaves the polynomial $X_1 + X_2 - X_3 - X_4$ fixed. Further find a polynomial fixed by the element (1234), but not by the whole group.

Exercise 4.17

Let G be a permutation group of degree n. Let p be a prime such that p does not divide n. Prove that any element of order p has to fix at least one point.

4.4 Counting Subgroups

We are almost ready to exploit group actions in counting arguments that involve geometry. But first some useful notation. If Y is a subset of X and H is a subset of G we write $H(Y) = \{h(y) : h \in H \text{ and } y \in Y\}$. If either H or Y consists of a single element we drop the set notation, so if $H = \{h\}$ we write $h(Y)$.

Definition 4.44 (Orbit of a group)

Let G act on X. An orbit of G is a subset $Y \subset X$ such that

(i) $G(Y) = Y$,

(ii) Y is non-empty and

(iii) if $\emptyset \not\subseteq U \not\subseteq Y$ then $G(U) \neq U$.

Lemma 4.45

If $x \in X$ then $G(x)$ is an orbit of G.

Proof

Let $y \in G(x)$. Then there is an element $h \in G$ so that $y = h(x)$ but now $g(y) = gh(x) = (gh)(x) \in G(x) \; \forall g \in G$. Clearly if $\emptyset \subsetneq U \subseteq G(x)$ and $u \in U$ then $u = k(x)$ for some $k \in G$. But now $x = k^{-1}k(x)$ so $U = G(x)$. $\qquad \square$

Lemma 4.46

Let G be a finite group and let G act on X. Define a relation, $\mathcal{R}$, on X by $x\mathcal{R}y$, if and only if there exists $g \in G$ with $g(x) = y$. Then $\mathcal{R}$ is an equivalence relation and the orbits are just the equivalence classes.

Proof

Let $x, y \in X$ with $x\mathcal{R}y$. Then there is an element $g \in G$ so that $y = g(x)$. So $g^{-1}(y) = x$. So $y\mathcal{R}x$. Now assume $y\mathcal{R}z$, then there exists $h \in G$ with $z = h(y)$. So $z = hg(x)$ and so $x\mathcal{R}z$. Since reflexivity is obvious we have that $\mathcal{R}$ is an equivalence relation. Note that $G(x)$ is just the equivalence class of x. $\qquad \square$

The next example is important.

Example 4.47

Let G be a group and H a subgroup of G. In this case we let X be the set of left cosets $\{xH : x \in G\}$. The action is defined by $xH \rightarrow gxH$.

We now prove a key result which relates orbits of elements to cosets of subgroups.

Theorem 4.48 (Orbit-Stabilizer Theorem)

Let G be a group and let G act on X. Let $x \in X$. Then there is a bijection between the elements of $G(x)$ and the cosets of G_x.

Proof

We need to define a map ψ from $G(x)$ to the cosets of G_x. Let $y \in G(x)$ and let $g \in G$ be such that $g(x) = y$, then define $\psi(y) = gG_x$. We have to show that ψ is well-defined and that it is a bijection.

To show that it is well-defined, consider $h \in G$ so that $h(x) = y = g(x)$. Then $h^{-1}g(x) = x$ so $h^{-1}g \in G_x$. So $hG_x = gG_x$ and ψ is well-defined.

To show that ψ is an injection assume $\psi(y) = \psi(z)$. So if $g(x) = y$ and $h(x) = z$

then $hG_x = gG_x$ and $h^{-1}g(x) = x$. But then $z = h(x) = g(x) = y$ and ψ is an injection. Finally we have to show that ψ is onto. Let gG_x be any coset of G_x and let $y = g(x)$ so then $\psi(y) = gG_x$. $\qquad\Box$

We now use this to give a proof of Theorem 4.6.

Proposition 4.49

Let G be a finite group and let H be a subgroup of G. Then the order of H divides the order of G.

Proof

We show that given any coset gH there is a bijection, ψ between gH and H. For any $k \in gH$ there is unique $h \in H$ with $k = gh$. Define $\psi(h) = k$. It is easy to see that ψ is well-defined and that it is a bijection. Since G is the disjoint union of the cosets of H and there are m cosets we see that $|G| = m|H|$. $\qquad\Box$

The number of cosets of a subgroup in a group is called the **index** and is written $[G : H]$. Hence, the previous theorem we write $|G| = |H|[G : H]$. This result, combined with the previous one, has far-reaching consequences, one of which is the **class equation** for finite groups.

Theorem 4.50 (Class equation for a finite group)

Let G be a finite group and let $C_1, C_2, \ldots C_k$ be the distinct conjugacy classes of G. Then $|C_i|$ divides $|G|$ for $1 \le i \le k$. Further

$$|G| = \sum_{1}^{k} |C_i|.$$

Proof

We note that the conjugacy classes are just the orbits of G acting on G by the conjugate action, Example 4.33. So by Theorem 4.48 and Proposition 4.49 we get the result. $\qquad\Box$

A simple corollary gives a well-known result about p-groups, which requires two definitions.

Definition 4.51 (A p-group)

Given a prime p, a **finite p-group** is a finite group whose order is a power of p.

Definition 4.52 (Centre of a group)

The centre of a group is the set of all elements which commute with all elements of the group: $\{z : zx = xz \,\forall x \in G\}$.

Note the identity is always in the centre so we say that the centre is **non-trivial** if it contains at least two elements.

Corollary 4.53

Let p be a prime and G be a finite p-group then the centre of G is non-trivial.

Proof

We observe that z is in the centre of G if and only if the conjugacy class of z is just $\{z\}$ and so has size 1. Let $|G| = p^a$. Now by the equation above

$$p^a = \sum_{1}^{k} |C_i|,$$

where $C_1, C_2, \ldots C_k$ are the distinct conjugacy classes of G. If we let C_1 be the conjugacy class that consists of the identity element then $|C_1| = 1$. Note, by the theorem above each $|C_i|$ divides p^a. So at least one other C_i has $|C_i| = 1$. So this conjugacy class consists of one non-identity element which is in the centre. $\square$

Another consequence is an alternative proof of Sylow's First Theorem.

Theorem 4.54 (Sylow's First Theorem)

Let G be a finite group of order $p^a m$ where p is a prime, a and m are natural numbers with $\gcd(p, m) = 1$. Then there exists a subgroup of G order p^a. Such a subgroup is called a Sylow p-subgroup.

Proof

Let X be the set of subsets of G of size p^a. There are $\binom{p^a m}{p^a}$ such subsets, Theorem 1.9 and such a number is coprime to p (see Exercise 4.26). Let G act on X by right multiplication. There must be a an orbit of size coprime to p. Let X be a subset in such an orbit and consider G_X the stabilizer of this subset.

Now $X G_X = X$ by the definition of G_X. So $X = \cup_{x \in X} x G_X$ and we can see that X is a union of cosets of G_X. So the order of G_X divides the size of X. So $|G_X|$ divides p^a. But $[G : G_X]$ is prime to p by Theorem 4.48. So p^a divides $|G_X|$ and hence $|G_X| = p^a$ and the theorem is proved.   □

4.4.1 Exercises

Exercise 4.18

Show that in Definition 4.30 for each $g \in G$ the mapping $\phi(g)$ is a permutation of X.

Exercise 4.19

Consider, as in Example 4.32, the action of G on G but define $\phi(g)(x) = xg^{-1}$. Prove that this is an action. Explain why the inverse is used.

Exercise 4.20

Let $g \in Sym(n)$. Consider the subgroup, say G, generated by g. Show that the cycles of g correspond to the orbits of G, that is if $(a_1, a_2, \ldots, a_r)$ is a cycle of g then $\{a_1, a_2, \ldots, a_r\}$ is an orbit of G.

Exercise 4.21

Calculate the order of the symmetry group of the cube acting on its vertices.

Exercise 4.22

Let G be a finite group of order n. By rewriting the class equation as

$$1 = \frac{1}{n} + \frac{n_1}{n} + \cdots + \frac{n_r}{n},$$

where there are $r + 1$ conjugacy classes of sizes $1, n_1, \ldots, n_r$, show that if there are only two conjugacy classes then $n = 2$. Further if there are only three conjugacy classes then $n = 6$.

4.5 Miscellaneous Exercises

Exercise 4.23

Let p be a prime and let G be a permutation group whose order is a power of p. Show that each orbit of G has to be a p-power.

Exercise 4.24

Let G act on a set X of size n. Let X_r be the set of sequences of elements of X of length r. Show that we can define an action on X_r by $g(x_1, x_2, x_3, \ldots, x_r) = (g(x_1), g(x_2), \ldots, g(x_r))$. (This is known as the diagonal action.)

Prove that, when G is the symmetric group action on $X = \{1, 2, \ldots, n\}$, there are two orbits of G on X_2 and five orbits on X_3. What is the general result?

Exercise 4.25

Let p be a prime and let n be a natural number. Show that the order of the Sylow p-subgroup of $\mathrm{Sym}(n)$ is given by

$$\sum_k \lfloor \tfrac{n}{p^k} \rfloor.$$

Exercise 4.26

Let p be a prime and a and b be natural numbers with $\gcd(p^a, m) = 1$. Prove that $\gcd\left(\binom{p^a m}{p^a}, p\right) = 1$.

Exercise 4.27

Show that the relationship of conjugacy is an equivalence relation.

5

Matrices, Sequences and Sums

5.1 Pascal's Triangle and Enumeration

This triangle – better described as an array – is associated with the Binomial Theorem 2.1; this expresses the powers of $(1+z)$ in terms of successive powers of z:

$$(1+z)^r = \sum_{k \geq 0} \binom{r}{k} z^k.$$

The next result shows that the coefficients involved represent the number of ways of choosing k elements from r, as assumed in the first chapter.

Theorem 5.1 (Choosing k objects from r)

The Binomial coefficient $\binom{r}{k}$ enumerates the number of ways of making a choice of k objects (without repetition) from r distinct objects.

Proof

We can write

$$(1+z)^r = \overbrace{(1+z)(1+z)\cdots(1+z)}^{r \text{ terms}}.$$

To form a term in z^k, on the right we must select a "z" from k brackets and there are r such brackets to choose from. So the resulting coefficient of z^k is simply the number of ways that this choice can be made. $\qquad\square$

A. Camina, B. Lewis, *An Introduction to Enumeration*,
Springer Undergraduate Mathematics Series,
DOI 10.1007/978-0-85729-600-9_5, © Springer-Verlag London Limited 2011

We are going to focus on the array made up from the Binomial coefficients. So our first step is to write the Binomial Theorem in terms of matrices

$$
\begin{pmatrix} 1 \\ 1+z \\ (1+z)^2 \\ (1+z)^3 \\ \cdots \end{pmatrix} = \begin{pmatrix} 1 & 0 & 0 & . & . \\ 1 & 1 & 0 & . & . \\ 1 & 2 & 1 & 0 & . \\ 1 & 3 & 3 & 1 & . \\ . & . & . & . & . \end{pmatrix} \begin{pmatrix} 1 \\ z \\ z^2 \\ z^3 \\ \cdots \end{pmatrix}
$$

and Pascal's triangle is the coefficient array that results. Pascal's triangle is generated by a recurrence relation.

Theorem 5.2 (Recurrence between terms of Pascal's array)

The recurrence relation satisfied by the entries in Pascal's array is

$$
\binom{r+1}{k} = \binom{r}{k} + \binom{r}{k-1}.
$$

Proof

Using the Binomial Theorem we can write

$$
(1+z)^{r+1} = \sum_{k \geq 0} \binom{r+1}{k} z^k = (1+z) \sum_{k \geq 0} \binom{r}{k} z^k
$$

and then comparing coefficients of z^k in each of the sums gives

$$
\binom{r+1}{k} = \binom{r}{k} + \binom{r}{k-1}
$$

exactly as required.   □

The terms that make up the array have many properties. An obvious property is the nature of the terms that make up each of the columns. For example, the first column consists of **units**; the non-zero terms in the second are simply **natural numbers** $\{1, 2, 3, \ldots\}$, while the non-zero terms in the third column are the **triangular numbers** $\{1, 3, 6, 10, \ldots\}$. The non-zero terms in the next column consist of the so-called **pyramidal numbers** $\{1, 4, 10, 20, 35, \ldots\}$. This sequence of sequences, which has terms that are partial sums of the sequence that precedes it, is the result of another basic property of the array.

Theorem 5.3 (Partial sums of columns of Pascal's array)

The partial sum of the terms in one column of Pascal's array is the term in the next column one below the bottom term of the partial sum:

$$\binom{r+1}{k+1} = \sum_{l=k}^{r} \binom{l}{k}.$$

Proof

Using the basic recurrence of Theorem 5.2 repeatedly we have

$$\binom{r+1}{k+1} = \binom{r}{k} + \binom{r}{k+1}$$

$$= \binom{r}{k} + \left(\binom{r-1}{k} + \binom{r-1}{k+1}\right)$$

$$\cdots$$

$$= \binom{r}{k} + \binom{r-1}{k} + \cdots + \binom{k}{k}$$

$$= \sum_{l=k}^{r} \binom{l}{k}.$$

as required. □

5.1.1 Exercises

Exercise 5.1

Prove the identity $\sum_{k \geq 0}(k+1)\binom{r}{k} = (r+2)2^{r-1}$.

Exercise 5.2

Prove by an enumerative argument that $k\binom{r}{k} = r\binom{r-1}{k-1}$.

Exercise 5.3

Prove, using an enumerative argument, that $\binom{r+1}{k} = \binom{r}{k} + \binom{r}{k-1}$. Deduce that the upward diagonal sums of terms in Pascal's triangle are Fibonacci numbers (Theorem 3.9).

Exercise 5.4

Show, using an enumerative argument, that $\binom{r}{l}\binom{r-l}{k-l} = \binom{r}{k}\binom{k}{l}$.

Exercise 5.5

Prove, using an enumerative argument, that $\binom{r}{k} = \sum_{l \geq 0}\binom{m}{l}\binom{r-m}{k-l}$ for $0 \leq m \leq r$.

5.2 The Push-Button Lock Sequence

We now investigate the problem of counting the number of push-button combinations there are on a push-button lock (such as those frequently used at the entrance to a building). We will restrict ourselves throughout to locks in which any button may be used only once.

Example 5.4

Imagine a four-button lock in which three buttons need to be pressed in the right order. How many such push-button lock combinations are there? If we exhaust all the possibilities, we find that there are 24 in all (Figure 5.1).

buttons 1, 2 and 3			buttons 1, 2 and 4		
(1 2 3)	(3 1 2)	(2 3 1)	(1 2 4)	(4 1 2)	(2 4 1)
(1 3 2)	(2 1 3)	(3 2 1)	(1 4 2)	(2 1 4)	(4 2 1)
buttons 2, 3 and 4			**buttons 1, 3 and 4**		
(2 3 4)	(4 2 3)	(3 4 2)	(1 3 4)	(4 1 3)	(3 4 1)
(2 4 3)	(3 2 4)	(4 3 2)	(1 4 3)	(3 1 4)	(4 3 1)

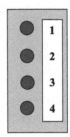

Figure 5.1 Lock combinations using three buttons from four.

Definition 5.5 (The push-button sequence $\{c_r\}$)

A push-button lock with r buttons (in which each button can be used only once) has c_r possible combinations. The push-button lock sequence is $\{c_r\}$.

Note: we count the empty combination (and it would fool most intruders!) so that $c_0 = 1$. This is consistent with previous definitions in which $\binom{0}{0} = 1 = 0!$.

Example 5.6

Next we find the total number of push-button lock combinations there are with four buttons. By the principle of exhaustion, a four-button lock uses 0, 1, 2, 3 or 4 buttons. If we use k buttons then we count as in Figure 5.2.

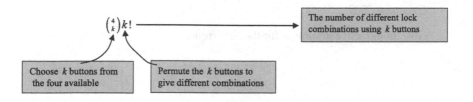

Figure 5.2 Lock combinations with k buttons.

In all then, the number of push-button lock combinations is

$$\sum_{k\geq 0}\binom{4}{k}k! = \binom{4}{0}0! + \binom{4}{1}1! + \binom{4}{2}2! + \binom{4}{3}3! + \binom{4}{4}4!$$

$$= 1 + 4 + 12 + 24 + 24 = 65.$$

An extension of this enumerative method proves the following result.

Theorem 5.7 (Explicit form for terms of the sequence $\{c_r\}$)

A push-button lock with r buttons (in which each button can be used only once) has

$$c_r = \sum_{k\geq 0}\binom{r}{k}k!$$

possible combinations and the sequence has the initial terms

$$\{c_r\} = \{1, 2, 5, 16, 65, 326, \ldots\}.$$

5.2.1 Exercises

Exercise 5.6

Prove, using the previous theorem and Exercise 5.2, that a recurrence satisfied by the push-button lock sequence $\{c_r\}$ is $c_r = rc_{r-1} + 1$ with the initial value $c_0 = 1$. Now prove this with an enumerative argument.

Exercise 5.7

Prove that another recurrence relation satisfied by the sequence of push-button locks is $c_{r+1} = (r+2)c_r - rc_{r-1}$.

Exercise 5.8

Give an inductive argument for the recurrence

$$r! = \sum_{k \geq 0} (-1)^{r-k} \binom{r}{k} c_k.$$

Exercise 5.9

The push-button lock sequence $\{c_r\}$ now satisfies the two indentities

(i) $c_r = \sum_{k \geq 0} \binom{r}{k} k!$;

(ii) $r! = \sum_{k \geq 0} (-1)^{r-k} \binom{r}{k} c_k$.

The derangement sequence $\{d_r\}$ also satisfies two identities. Each identity is an adaptation of Example 1.27 and Equation (1.2), respectively,

(i) $d_r = \sum_{k \geq 0} (-1)^{r-k} \binom{r}{k} k!$;

(ii) $r! = \sum_{k \geq 0} \binom{r}{k} d_k$.

Write each of these four expressions in matrix form.

5.3 Pascal's Triangle as a Matrix

Once we begin to think of the array as a matrix, a formidable problem arises: the array is unbounded. Surprisingly, this causes no problems. Pascal's triangle begins like this:

$$P = \begin{pmatrix} 1 & 0 & 0 & . & . \\ 1 & 1 & 0 & . & . \\ 1 & 2 & 1 & 0 & . \\ 1 & 3 & 3 & 1 & . \\ . & . & . & . & . \end{pmatrix}$$

Definition 5.8 (Pascal's triangle as a matrix)

Pascal's triangle is the infinite matrix $P = \left(\binom{r}{k} : r, k \geq 0 \right)$. Notice that we number rows and columns from 0 not 1.

Having acknowledged that we are dealing with an infinite matrix, a matrix that extends indefinitely downwards and to the right, we also note that it has a special form.

Definition 5.9 (Lower triangular matrix)

A lower triangular matrix $M = (m_{r,k})$ is one in which $m_{r,k} = 0$ if $k > r$.

Every lower triangular matrix (finite or infinite) has well-defined positive powers. This is simply the result of its lower triangular structure and an important property: every row of a lower triangular matrix has only a finite number of non-zero terms. Hence the **product** of two lower triangular matrices is itself lower triangular. This means that such multiplication – and hence powers – is always well-defined.

Negative powers can also be defined for such arrays whenever an inverse exists. Fortunately there is a simple condition that ensures this.

Lemma 5.10 (Determinant of a lower triangular matrix)

The determinant of a finite square lower triangular matrix is the product of the entries on its leading diagonal.

Proof

We use induction. To seed this

$$\left| \begin{pmatrix} m_{0,0} & \\ m_{1,0} & m_{1,1} \end{pmatrix} \right| = m_{0,0} m_{1,1}$$

and so it is true for 2×2 lower triangular matrices. Now suppose the $n \times n$ lower triangular matrix $(m_{r,k})$ has the determinant $m_{0,0}.m_{1,1}. \cdots m_{n,n}$. Then the $n+1 \times n+1$ lower triangular matrix $(m_{r,k})$ has the determinant:

$$\left| \begin{pmatrix} m_{0,0} & & & \\ m_{1,0} & m_{1,1} & & \\ \cdot & \cdot & & \\ \cdot & \cdot & \cdot & \\ m_{n+1,0} & & & m_{n+1,n+1} \end{pmatrix} \right| = m_{0,0} \left| \begin{pmatrix} m_{1,1} & & & \\ m_{2,1} & m_{2,2} & & \\ \cdot & \cdot & & \\ \cdot & \cdot & \cdot & \\ m_{n+1,0} & & & m_{n+1,n+1} \end{pmatrix} \right|$$

which, by hypothesis, is simply $m_{0,0}(m_{1,1}. \cdots m_{n+1,n+1})$, exactly as required for the induction to succeed. □

This means that so long as there are no zeros on the leading diagonal, the (finite) inverse exists and this leads directly to the inverse of the infinite array. In turn this leads to any negative power of the array. We investigate the powers of Pascal's triangle.

Theorem 5.11 (Inverse of Pascal's matrix)

The inverse of Pascal's triangle is given by $P^{-1} = \left((-1)^{r-k} \binom{r}{k} \right)$.

Proof

The Binomial expansion

$$(1+z)^r = \sum_{k \geq 0} \binom{r}{k} z^k \tag{5.1}$$

expresses powers of $(1+z)$ in terms of powers of z. To find the inverse of Pascal's triangle we must invert this equation: that is, to express powers of z in terms of powers of $(1+z)$. The resulting coefficients will form an array that is the inverse of the Pascal array. But Equation (5.1) may be transformed by $z \to -z - 1$ and becomes

$$(-1)^r z^r = \sum_{l \geq 0} (-1)^l \binom{r}{l} (1+z)^l.$$

This may be rewritten:

$$z^r = \sum_{k \geq 0} (-1)^{r-k} \binom{r}{k} (1+z)^k \tag{5.2}$$

Equation (5.2) is the required inverted form of Equation (5.1), and we conclude that $P^{-1} = \left((-1)^{r-k} \binom{r}{k} \right)$. $\qquad\square$

The inverse has the same terms as Pascal's matrix but with alternating signs – positive if the indices r and k have the same parity and negative otherwise:

$$P^{-1} = \begin{pmatrix} 1 & 0 & 0 & . & . \\ -1 & 1 & 0 & . & . \\ 1 & -2 & 1 & 0 & . \\ -1 & 3 & -3 & 1 & . \\ . & . & . & . & . \end{pmatrix}.$$

This means that the inverse of Pascal's array is a signed copy of itself. Now look back at Exercise 5.9. We can make sense of the four identities there – they involve Pascal's triangle and its inverse. The derangement numbers $\{d_r\}$ satisfy the recurrence

$$r! = \sum_{k \geq 0} \binom{r}{k} d_k$$

which may be written in the matrix form

$$\begin{pmatrix} 0! \\ 1! \\ 2! \\ 3! \\ \cdots \end{pmatrix} = \begin{pmatrix} 1 & 0 & 0 & . & . \\ 1 & 1 & 0 & . & . \\ 1 & 2 & 1 & 0 & . \\ 1 & 3 & 3 & 1 & . \\ . & . & . & . & . \end{pmatrix} \begin{pmatrix} d_0 \\ d_1 \\ d_2 \\ d_3 \\ \cdots \end{pmatrix}.$$

If we multiply each side by the inverse matrix P^{-1} we have

$$
\begin{pmatrix} d_0 \\ d_1 \\ d_2 \\ d_3 \\ \cdots \end{pmatrix} = P^{-1} \begin{pmatrix} 0! \\ 1! \\ 2! \\ 3! \\ \cdots \end{pmatrix} = \begin{pmatrix} 1 & 0 & 0 & \cdot & \cdot \\ -1 & 1 & 0 & \cdot & \cdot \\ 1 & -2 & 1 & 0 & \cdot \\ -1 & 3 & -3 & 1 & \cdot \\ \cdot & \cdot & \cdot & \cdot & \cdot \end{pmatrix} \begin{pmatrix} 0! \\ 1! \\ 2! \\ 3! \\ \cdots \end{pmatrix}
$$

which is, of course,

$$
d_r = \sum_{k \geq 0} (-1)^{r-k} \binom{r}{k} k!
$$

the explicit form for the derangement numbers obtained independently.

An identical pair of relationships is true for the push-button lock sequence:

$$
c_r = \sum_{k \geq 0} \binom{r}{k} k!
$$

which has the matrix form

$$
\begin{pmatrix} c_0 \\ c_1 \\ c_2 \\ c_3 \\ \cdots \end{pmatrix} = \begin{pmatrix} 1 & 0 & 0 & \cdot & \cdot \\ 1 & 1 & 0 & \cdot & \cdot \\ 1 & 2 & 1 & 0 & \cdot \\ 1 & 3 & 3 & 1 & \cdot \\ \cdot & \cdot & \cdot & \cdot & \cdot \end{pmatrix} \begin{pmatrix} 0! \\ 1! \\ 2! \\ 3! \\ \cdots \end{pmatrix}
$$

while the recurrence

$$
r! = \sum_{k \geq 0} (-1)^{r-k} \binom{r}{k} c_k
$$

has the matrix form

$$
\begin{pmatrix} 0! \\ 1! \\ 2! \\ 3! \\ \cdots \end{pmatrix} = \begin{pmatrix} 1 & 0 & 0 & \cdot & \cdot \\ -1 & 1 & 0 & \cdot & \cdot \\ 1 & -2 & 1 & 0 & \cdot \\ -1 & 3 & -3 & 1 & \cdot \\ \cdot & \cdot & \cdot & \cdot & \cdot \end{pmatrix} \begin{pmatrix} c_0 \\ c_1 \\ c_2 \\ c_3 \\ \cdots \end{pmatrix}.
$$

These are two examples of a general principle.

Theorem 5.12 (Inverting a pair of sequences related by Pascal)

Suppose that a pair of sequences $\{u_r\}$ and $\{v_r\}$ are such that

$$
v_r = \sum_{k \geq o} \binom{r}{k} u_k,
$$

then we can invert this sum

$$u_r = \sum_{k \geq 0} (-1)^{r-k} \binom{r}{k} v_k.$$

Example 5.13

We can use this to illustrate a surprising property, and the inverse, of the Fibonacci numbers that $F_{2r} = \sum_{k \geq 0} \binom{r}{k} F_k$. Using Binet's formula we have

$$\sum_{k \geq 0} \binom{r}{k} F_k = \sum_{k \geq 0} \binom{r}{k} \frac{1}{\sqrt{5}} \left(\phi_1^k - \phi_2^k \right) = \frac{1}{\sqrt{5}} \left(\sum_{k \geq 0} \binom{r}{k} \phi_1^k - \sum_{k \geq 0} \binom{r}{k} \phi_2^k \right).$$

Each sum inside the bracket is part of a Binomial expansion, and hence

$$\sum_{k \geq 0} \binom{r}{k} F_k = \frac{1}{\sqrt{5}} \left((1 + \phi_1)^r - (1 + \phi_2)^r \right) = \frac{1}{\sqrt{5}} \left((\phi_1^2)^r - (\phi_2^2)^r \right)$$

since $\phi_i^2 = \phi_i + 1$ for $i = 1, 2$, see Exercise 3.16; this means that

$$\sum_{k \geq 0} \binom{r}{k} F_k = \frac{1}{\sqrt{5}} \left(\phi_1^{2r} - \phi_2^{2r} \right) = F_{2r}.$$

Using the previous theorem we can "invert" this sum and then we have the "dual" result

$$F_r = \sum_{k \geq 0} (-1)^{r-k} \binom{r}{k} F_{2k}.$$

We continue our investigation of the powers of Pascal's triangle.

Theorem 5.14 (The square of Pascal's triangle)

The square of Pascal's triangle is given by $P^2 = 2^{r-k} \binom{r}{k}$.

Proof

Again we manipulate the Binomial Theorem in two different ways and compare the results. First

$$(2 + z)^r = 2^r (1 + \frac{z}{2})^r = 2^r \sum_{k \geq 0} \binom{r}{k} \frac{z^k}{2^k} = \sum_{k \geq 0} 2^{r-k} \binom{r}{k} z^k. \tag{5.3}$$

But we also have

$$(2+z)^r = (1+(1+z))^r = \sum_{l \geq 0} \binom{r}{l}(1+z)^l = \sum_{l \geq 0} \binom{r}{l} \sum_{k \geq 0} \binom{l}{k} z^k$$

using the Binomial Theorem twice. The repeated sums are both finite, so we may reverse them:

$$(2+z)^r = \sum_{k \geq 0} \left(\sum_{l \geq 0} \binom{r}{l}\binom{l}{k} \right) z^k. \tag{5.4}$$

Comparing the coefficients of z^k in Equations (5.3) and (5.4) leads to

$$\sum_{l \geq 0} \binom{r}{l}\binom{l}{k} = 2^{r-k}\binom{r}{k}$$

which is precisely the statement that $P.P = P^2 = \left(2^{r-k}\binom{r}{k} \right)$ as required. $\qquad\square$

So the square of Pascal starts like this

$$P^2 = \begin{pmatrix} 1 & 0 & 0 & . & . \\ 2.1 & 1 & 0 & . & . \\ 2^2.1 & 2.2 & 1 & 0 & . \\ 2^3.1 & 2^2.3 & 2.3 & 1 & . \\ . & . & . & . & . \end{pmatrix}$$

These results (based on the limited evidence of $n = 1$, $n = -1$ and $n = 2$) seem to suggest a general result for powers of Pascal's matrix: that $P^n = \left(n^{r-k}\binom{r}{k} \right)$.

Theorem 5.15 (Powers of Pascal's matrix)

The integer powers of Pascal's array (for $n \in \mathbb{Z}$) are given by $P^n = \left(n^{r-k}\binom{r}{k} \right)$ where we interpret $P^0 = I$, the identity matrix.

Proof (For n positive)

We use induction. We have already established it for a positive "seed". Suppose it is true for $n - 1$ so that $P^{n-1} = \left((n-1)^{r-k}\binom{r}{k} \right)$. Now

$$(n+z)^r = (1+(n-1+z))^r = \sum_{l \geq 0} \binom{r}{l}(n-1+z)^l = \sum_{l \geq 0} \binom{r}{l} \sum_{k \geq 0} \binom{l}{k}(n-1)^{l-k} z^k$$

$$= \sum_{k \geq 0} \left(\sum_{l \geq 0} \binom{r}{l}(n-1)^{l-k}\binom{l}{k} \right) z^k.$$

The inner sum is simply the terms of the matrix product $P.P^{n-1}$. On the other hand

$$(n+z)^r = \sum_{k \geq 0} n^{r-k} \binom{r}{k} z^k.$$

Comparing the coefficients of z^k in each case establishes the result required. This proof is easily adapted to the case when n is negative.   □

This proves another general result about sequences related through powers of Pascal.

Theorem 5.16 (Pairs of sequences related by powers of Pascal)

Suppose that a pair of sequences $\{u_r\}$ and $\{v_r\}$ are such that

$$v_r = \sum_{k \geq 0} n^{r-k} \binom{r}{k} u_k,$$

where $n \in \mathbb{Z}$. Then we can invert this sum

$$u_r = \sum_{k \geq 0} (-n)^{r-k} \binom{r}{k} v_k.$$

5.3.1 Exercises

Exercise 5.10

We may extend Theorem 5.15 from integers to rational powers; suppose we consider the square root of Pascal's array – that is a matrix whose square is just the Pascal array. We might suppose that $\sqrt{P} = \left(2^{k-r} \binom{r}{k} \right)$ so that,

$$P^{\frac{1}{2}} = \begin{pmatrix} 1 & & & & & \cdot \\ \frac{1}{2} & 1 & & & & \cdot \\ \frac{1}{4} & 1 & 1 & & & \cdot \\ \frac{1}{8} & \frac{3}{4} & \frac{3}{2} & 1 & & \cdot \\ \frac{1}{16} & \frac{4}{8} & \frac{6}{4} & \frac{4}{2} & 1 & \cdot \\ \cdot & \cdot & \cdot & \cdot & \cdot & \cdot \end{pmatrix}.$$

Verify that (for initial rows and columns) $P^{\frac{1}{2}}.P^{\frac{1}{2}} = P$.

Exercise 5.11

Show that the finite Pascal matrix

$$P_3 = \begin{pmatrix} 1 & 0 & 0 \\ 1 & 1 & 0 \\ 1 & 2 & 1 \end{pmatrix}$$

satisfies the equation $(I - P_3)^3 = 0$ where I is the 3×3 identity matrix. Generalize.

Exercise 5.12

A sequence $\{B_r\}$ obeys the recurrence

$$B_{r+1} = \sum_{k \geq 0} \binom{r}{k} B_k$$

which is "seeded" by the initial value $B_0 = 1$. (This sequence is called the **Bell sequence** and it counts the number of ways of partitioning a set into non-empty, pairwise disjoint, subsets. We will meet it again in Chapter 7, Example 7.14.

(i) Use the recurrence to calculate the next four terms of the sequence;

(ii) prove that $B_r = \sum_{k \geq 0} (-1)^{r-k} \binom{r}{k} B_{k+1}$.

5.4 Operations on Generating Functions

In the three-step recipe (Algorithm 2.23) to convert recurrences into generating functions, most of the action takes place in the final step. Success is determined by the way that components of the recurrence can be written in terms of the desired generating function and other known functions. In this section we concentrate on "doing" things to generating functions to see what results – the expectation is that later we may reverse this process to convert difficult terms in a recurrence into generating functions. At the same time we explore the way that "doing" things to a given generating function changes the underlying sequence.

5.4.1 Multiplication by z^k

Suppose we start with a given sequence $\{u_r\}$ with generating function, $U(z)$. If we multiply this by z^k then we have $z^k U(z) = \sum_{r \geq 0} u_r z^{r+k}$ which is the generating function for the sequence,

$$\{\overbrace{0, 0, \ldots, 0}^{k \text{ zeros}}, u_0, u_1, \ldots\}.$$

We used this idea in Example 2.22.

5.4.2 The Product of Two Generating Functions

Consider the product of two generating function $W(z) = U(z)V(z)$. The function $W(z)$ is itself a generating function, but for what sequence? It turns out that the answer is simple and has powerful and varied applications.

Theorem 5.17 (The generating function product formula)

If $W(z) = U(z)V(z)$, where U, V and W are the generating functions for the sequences $\{u_r\}, \{v_r\}$ and $\{w_r\}$, then

$$w_r = \sum_{k=0}^{r} u_k v_{r-k} = \sum_{k=0}^{r} u_{r-k} v_k.$$

Proof

The product of the two generating functions may be written out in the form:

$$\left(u_0 + u_1 z + u_2 z^2 + \cdots + u_r z^r + \cdots\right) \left(v_0 + v_1 z + v_2 z^2 + \cdots + v_r z^r + \cdots\right).$$

If we want the coefficient of z^r in this product, then we make it up from the sum of the products of pairs of terms taken one from each bracket. Moreover, the sum of the corresponding powers of z, of which these pairs are the coefficients, must be r. So we have:

$$w_r = \sum_{k+l=r} u_k v_l$$

We may rewrite this in two (equivalent) forms

$$w_r = \sum_{k=0}^{r} u_k v_{r-k} = \sum_{k=0}^{r} u_{r-k} v_k.$$

$\square$

Example 5.18

We use the product of the Fibonacci and Lucas generating functions to derive an identity. We may write the generating functions for the Fibonacci and Lucas sequences in the forms,

$$\sum_{r \geq 0} F_r z^r = \frac{1}{\sqrt{5}} \left(\frac{1}{1 - \phi_1 z} - \frac{1}{1 - \phi_2 z} \right) \text{ and } \sum_{r \geq 0} L_r z^r = \left(\frac{1}{1 - \phi_1 z} + \frac{1}{1 - \phi_2 z} \right).$$

(Using these forms of the generating functions, means fewer partial fractions later.) Now we take the product of these generating functions and then

$$\left(\sum_{r\geq0} F_r z^r\right)\left(\sum_{r\geq0} L_r z^r\right) = \frac{1}{\sqrt{5}}\left(\frac{1}{(1-\phi_1 z)^2} - \frac{1}{(1-\phi_2 z)^2}\right).$$

Expand each term on the right,

$$\left(\sum_{r\geq0} F_r z^r\right)\left(\sum_{r\geq0} L_r z^r\right) = \frac{1}{\sqrt{5}}\sum_{r\geq0}(r+1)\left(\phi_1^r - \phi_2^r\right)z^r = \sum_{r\geq0}(r+1)F_r z^r.$$

We deduce, by the theorem above, that

$$\sum_{k=0}^{r} F_k L_{r-k} = (r+1)F_r.$$

5.4.3 Partial Sums of a Sequence

If we are given a sequence $\{u_r\}$ it is of interest to find the sequence $\{v_r\}$ which has as its terms the partial sums $v_r = \sum_{k=0}^{r} u_k$.

Theorem 5.19 (The sum formula)

Let $\{u_r\}$ be a sequence with a generating function $U(z)$ then a generating function for the sequence $\{v_r\}$ that consists of partial sums of the sequence $\{u_r\}$ is given by the product $\frac{1}{1-z}U(z)$.

Proof

This is a simple consequence of the generating function product formula, Theorem 5.17. If we write $V(z) = \frac{1}{1-z}U(z)$ and then write these as power series, we have

$$\sum_{r\geq0} v_r z^r = \left(\sum_{r\geq0} 1 z^r\right)\left(\sum_{r\geq0} u_r z^r\right)$$

and then

$$v_r = \sum_{k=0}^{r} 1.u_k = \sum_{k=0}^{r} u_k$$

exactly as required. $\qquad\square$

The same idea can result in a different form of the partial sum – a "weighted" sum. If we multiply a generating function by a function that resembles $\frac{1}{1-z}$ then the result is still a sum, but with a "weighting" according to the nature of the particular function used.

Example 5.20

We take the product of the generating function of the Lucas numbers with the generating function $\frac{1}{2-z}$ to prove that $\sum\limits_{k=0}^{r} 2^k L_k = 2^{r+1} F_{r+1}$. The generating function of the Lucas numbers is given by $\sum_{r\geq 0} L_r z^r = \frac{2-z}{1-z-z^2}$: if we multiply by the reciprocal of the numerator on the right we have

$$\frac{1}{2-z}\sum_{r\geq 0} L_r z^r = \frac{1}{1-z-z^2} = \frac{1}{z}\left(\frac{z}{1-z-z^2}\right) = \frac{1}{z}\sum_{r\geq 0} F_r z^r;$$

a slight rearrangement gives

$$\frac{1}{\left(1-\frac{z}{2}\right)}\sum_{r\geq 0} L_r z^r = \frac{2}{z}\sum_{r\geq 0} F_r z^r.$$

The product on the left becomes

$$\left(\sum_{r\geq 0}\frac{z^r}{2^r}\right)\left(\sum_{r\geq 0} L_r z^r\right) = \frac{2}{z}\sum_{r\geq 0} F_r z^r$$

and then comparing coefficients of z^r on either side (making use of the generating function product formula, Theorem 5.17) gives

$$\sum_{k=0}^{r}\frac{L_k}{2^{r-k}} = 2F_{r+1}$$

and hence

$$\sum_{k=0}^{r} 2^k L_k = 2^{r+1} F_{r+1}.$$

There are even more exotic ways that we can exploit this basic idea.

Example 5.21

We start with two given generating functions for $\{F_r^2\}$ and $\{F_{2r}\}$,

$$\sum_{r\geq 0} F_r^2 z^r = \frac{z(1-z)}{(1+z)(1-3z+z^2)} \quad \text{and} \quad \sum_{r\geq 0} F_{2r} z^r = \frac{z}{1-3z+z^2}.$$

The first generating function has the factor $1 - z$ in the numerator, so we divide both sides by it

$$\left(\frac{1}{1-z}\right) \sum_{r \geq 0} F_r^2 z^r = \frac{z}{(1+z)(1-3z+z^2)}.$$

Partial fractions now give

$$\left(\frac{1}{1-z}\right)\left(\sum_{r \geq 0} F_r^2 z^r\right) = \frac{1}{5}\left(\frac{1+z}{1-3z+z^2} - \frac{1}{1+z}\right)$$

$$= \frac{1}{5}\left(\frac{1}{z}\left(\frac{z}{1-3z+z^2}\right) + \frac{z}{1-3z+z^2} - \frac{1}{1+z}\right)$$

$$= \frac{1}{5}\left(\frac{1}{z}\sum_{r \geq 0} F_{2r}z^r + \sum_{r \geq 0} F_{2r}z^r - \sum_{r \geq 0} (-1)^r z^r\right).$$

Using the Sum formula, Theorem 5.19, provides the interesting identity

$$\sum_{k=0}^{r} F_k^2 = \frac{1}{5}\left(F_{2r+2} + F_{2r} - (-1)^r\right).$$

5.4.4 The $zD \equiv z\frac{d}{dz}$ Operation

We have already made use of differentiation (Example 2.21). Now we look at repeated differentiation with an added ingredient.

Definition 5.22 (The Euler operation)

The zD operation (differentiate a function and then multiply by the argument) is called the Euler operation.

We apply the Euler operation to the function $\frac{1}{1-z}$. We may expand the function as a power series (when $|z| < 1$)

$$f(z) = \sum_{r \geq 0} z^r = \frac{1}{1-z}.$$

A first application of the Euler operation gives

$$(zD)f(z) = \sum_{r \geq 0} rz^r = \frac{z}{(1-z)^2}$$

(We may differentiate term by term inside the disc of convergence of a power series.)
Using the operation again gives

$$(zD)^2 f(z) = \sum_{r\geq0} r^2 z^r = z \cdot \frac{(1-z)^2 + z.2.(1-z)}{(1-z)^4} = \frac{z+z^2}{(1-z)^3}.$$

Repeating the same operation one more time

$$(zD)^3 f(z) = \sum_{r\geq0} r^3 z^r = z \cdot \frac{(1+2z)(1-z)^3 + (z+z^2).3.(1-z)^2}{(1-z)^6} = \frac{z+4z^2+z^3}{(1-z)^4}.$$

$$(5.5)$$

The coefficients appearing in the numerators follow a pattern that may be written in
the form of an array which continues:

$$\begin{pmatrix} 1 & & & & & & \cdot \\ 0 & 1 & & & & & \cdot \\ 0 & 1 & 1 & & & & \cdot \\ 0 & 1 & 4 & 1 & & & \cdot \\ 0 & 1 & 11 & 11 & 1 & & \cdot \\ 0 & 1 & 26 & 66 & 26 & 1 & \cdot \\ \cdot & \cdot & \cdot & \cdot & \cdot & \cdot & \cdot \end{pmatrix}.$$

Definition 5.23 (The Eulerian array)

The coefficients that appear in the numerators at each stage of the application of the
Euler operation to the function $\frac{1}{1-z}$ are called Eulerian numbers and make up an array
called the Eulerian array. We write $E = \left(\left\langle \begin{smallmatrix} r \\ k \end{smallmatrix} \right\rangle\right)$.

Note: The symbol chosen to represent the Eulerian numbers mimics those used for
Binomial coefficients. The indices representing the columns (k) and rows (r) are each
numbered from zero; for example, $\left\langle \begin{smallmatrix} 0 \\ 0 \end{smallmatrix} \right\rangle = 1$, $\left\langle \begin{smallmatrix} 3 \\ 2 \end{smallmatrix} \right\rangle = 4$.

Theorem 5.24

The Eulerian numbers obey the recurrence

$$\left\langle \begin{matrix} n+1 \\ k \end{matrix} \right\rangle = (n-k+2)\left\langle \begin{matrix} n \\ k-1 \end{matrix} \right\rangle + k\left\langle \begin{matrix} n \\ k \end{matrix} \right\rangle.$$

Proof

From the definition

$$(zD)^n (f(z)) = \sum_{r \geq 0} r^n z^r = \frac{\sum_{k \geq 0} \left\langle {n \atop k} \right\rangle z^k}{(1-z)^{n+1}}.$$

If we apply the Euler operation we have

$$(zD)^{n+1} (f(z)) = (zD) \left(\frac{\sum_{k \geq 0} \left\langle {n \atop k} \right\rangle z^k}{(1-z)^{n+1}} \right)$$

But this may be written as

$$\frac{\sum_{k \geq 0} \left\langle {n+1 \atop k} \right\rangle z^k}{(1-z)^{n+2}} = (zD) \left(\frac{\sum_{k \geq 0} \left\langle {n \atop k} \right\rangle z^k}{(1-z)^{n+1}} \right)$$

and carrying out the Euler operation on the right

$$\frac{\sum_{k \geq 0} \left\langle {n+1 \atop k} \right\rangle z^k}{(1-z)^{n+2}} = z \left(\frac{(1-z)^{n+1} \sum_{k \geq 0} k \left\langle {n \atop k} \right\rangle z^{k-1} + (n+1)(1-z)^n \sum_{k \geq 0} \left\langle {n \atop k} \right\rangle z^k}{(1-z)^{2n+2}} \right)$$

which can be simplified.

$$\frac{\sum_{k \geq 0} \left\langle {n+1 \atop k} \right\rangle z^k}{(1-z)^{n+2}} = z \left(\frac{(1-z) \sum_{k \geq 0} k \left\langle {n \atop k} \right\rangle z^{k-1} + (n+1) \sum_{k \geq 0} \left\langle {n \atop k} \right\rangle z^k}{(1-z)^{n+2}} \right).$$

Next remove the common denominators so that

$$\sum_{k \geq 0} \left\langle {n+1 \atop k} \right\rangle z^k = z(1-z) \sum_{k \geq 0} k \left\langle {n \atop k} \right\rangle z^{k-1} + z(n+1) \sum_{k \geq 0} \left\langle {n \atop k} \right\rangle z^k$$

and now we may compare coefficients of z^k on either side to give

$$\left\langle {n+1 \atop k} \right\rangle = k \left\langle {n \atop k} \right\rangle - (k-1) \left\langle {n \atop k-1} \right\rangle + (n+1) \left\langle {n \atop k-1} \right\rangle$$

$$= (n-k+2) \left\langle {n \atop k-1} \right\rangle + k \left\langle {n \atop k} \right\rangle,$$

as required. $\square$

This recurrence leads directly to an enumerative role for the Eulerian numbers. Rather amazingly, given their origins, they occur in counting the number of **ascents** in a permutation.

Definition 5.25 (Ascents in a permutation)

Let π be any permutation of $\{1, 2, \ldots r\}$; an ascent occurs whenever $\pi(i) < \pi(i+1)$.

Example 5.26

Given a particular permutation it is easy to count the number of ascents it has. The permutation $(13)(24)$ has the bijective form:

$$(1\ 3)(2\ 4) \equiv \begin{pmatrix} 1\ 2\ 3\ 4 \\ 3\ 4\ 1\ 2 \end{pmatrix}.$$

The image of this permutation is (3412) and it has two ascents: $3 \nearrow 4$ and $1 \nearrow 2$.

Definition 5.27 (Counting ascents in a permutation)

We denote by $a_{r,k}$ the number of permutations of r objects that have precisely k ascents.

The ascent numbers $a_{r,k}$ make up an array.

Example 5.28

We will find the entries in the row corresponding to $r = 3$, that is, the distribution of ascents in the permutations of $\{1, 2, 3\}$. First we list them in cycle form (since it is easy to list them in this form) and then the image of the corresponding bijective form; from the latter we can easily read off the number of ascents (Table 5.1).

We conclude that of the permutations of $\{1, 2, 3\}$ there is (are):

 (i) **1** permutation with no ascents;

 (ii) **4** permutations with a single ascent;

(iii) **1** permutation with two ascents.

So $a_{3,0} = 1$, $a_{3,1} = 4$, $a_{3,2} = 1$ and $a_{3,3} = 0$.

These results are very suggestive. We shall prove that the distribution of ascents in permutations is closely related to the terms of the Eulerian array. We do this by finding a recurrence they satisfy.

Cycle form of permutation	Image	Number of ascents
(1)(2)(3)	1 2 3	2
(1 2)(3)	2 1 3	1
(1 3)(2)	3 2 1	0
(2 3)(1)	1 3 2	1
(1 2 3)	2 3 1	1
(1 3 2)	3 1 2	1

Table 5.1 Ascents in permutations of $\{1,2,3\}$.

Theorem 5.29 (Recurrence for ascent numbers)

The recurrence satisfied by the ascent numbers is:

$$a_{r,k} = (k+1)a_{r-1,k} + (r-k)a_{r-1,k-1}.$$

Proof

Each permutation σ of $\{1,2,\ldots,r-1\}$ leads to r permutations of $\{1,2,\ldots,r-1,r\}$ if we insert the new element r in all possible ways. Suppose we put r in position l then we obtain the permutation

$$\pi = \sigma(1)\cdots\sigma(l-1)r\sigma(l)\cdots\sigma(r-1).$$

The number of ascents in π is the same as the number in σ if $l=1$ or $\sigma(l-1)<\sigma(l)$; the number of ascents in π is greater than the number in σ if $\sigma(l-1)>\sigma(l)$ or if $l=r$. These are the only possibilities.

So π has k ascents in a total of $(k+1)a_{r-1,k}$ ways from permutations σ that have k ascents, plus a total of $((r-2)-(k-1)+1)a_{r-1,k-1}$ ways from permutations σ that have $k-1$ ascents. Hence

$$a_{r,k} = (k+1)a_{r-1,k} + (r-k)a_{r-1,k-1}.$$

$\square$

Theorem 5.30 (The ascent numbers and Eulerian numbers)

The number of permutations of $\{1,2,\cdots,r\}$ with k ascents is given by $a_{0,0}=1$ and $a_{r,k} = \left\langle {r \atop k+1} \right\rangle$ for $r \geq 1$.

Proof

The Eulerian numbers obey the recurrence

$$\left\langle {r \atop k+1} \right\rangle = (r-k)\left\langle {r-1 \atop k} \right\rangle + (k+1)\left\langle {r-1 \atop k+1} \right\rangle$$

which is exactly the same recurrence as that for $a_{r,k}$. So the Eulerian numbers and the ascent numbers $a_{r,k}$ share the same initial values and the same recurrence – they take identical values, suitably displaced.   □

Corollary 5.31 (Properties of Eulerian and ascent numbers)

The Eulerian numbers have the property that

$$\sum_{k\geq 0} \left\langle {r \atop k} \right\rangle = r! \quad \text{and} \quad \left\langle {r \atop k} \right\rangle = \left\langle {r \atop r-k+1} \right\rangle.$$

Proof

The first part is an easy consequence of the fact that $\sum_{k\geq 0} a_{r,k} = r!$ because they count all the permutations of $\{1,2,\ldots,r\}$ since any permutation *either* has no ascents, *or* 1 ascent etc. The second comes from another enumerative insight. There are as many permutations of r objects with k ascents as there are with k descents. Moreover, if a permutation of r objects has k ascents then it has $r-k$ descents; hence the result.   □

Rather surprisingly, we can use the Euler operation to obtain sums of powers of positive integers.

Example 5.32

We seek a formula for the sum of the first n cubes of the positive integers $\sum_{k=1}^{n} k^3$. We need only use the sum operation of Theorem 5.19 on Equation (5.5) when n takes the value $n = 3$. We have

$$\left(\frac{1}{1-z}\right)\sum_{r\geq 0} r^3 z^r = \left(\frac{1}{1-z}\right)\frac{\sum_{k\geq 0} \left\langle {3 \atop k} \right\rangle z^k}{(1-z)^4} = \frac{z+4z^2+z^3}{(1-z)^5} = (z+4z^2+z^3)\sum_{r\geq 0} \binom{r+4}{4} z^r.$$

Comparing the coefficients of z^n on either side we have

$$\sum_{k=1}^{n} k^3 = \binom{n+3}{4} + 4\binom{n+2}{4} + \binom{n+1}{4}.$$

So

$$\sum_{k=1}^{n} k^3 = \frac{n^2(n+1)^2}{4}.$$

In principle, we have solved the problem of finding the sums of the powers of the positive integers for any positive integer power. We simply make use of the corresponding row of the Eulerian array. In fact we can extend this further. The corresponding result for the alternating sum comes when we apply an additional twist. We start with a sibling function

$$\sum_{r\geq0}(-1)^r z^r = \frac{1}{1+z};$$

a first application of the Euler operation gives

$$\sum_{r\geq0}r(-1)^r z^r = \frac{-z}{(1+z)^2};$$

using the operation yet again gives,

$$\sum_{r\geq0}r^2(-1)^r z^r = z.\frac{-(1+z)^2+z.2.(1+z)}{(1+z)^4} = \frac{-z+z^2}{(1+z)^3}.$$

Repeating the same operation one more time

$$\sum_{r\geq0}r^3(-1)^r z^r = z.\frac{(-1+2z)(1+z)^3-(-z+z^2).3.(1+z)^2}{(1+z)^6}$$

$$= \frac{-z+4z^2-z^3}{(1+z)^4}.$$

Notice the appearance of the signed Eulerian array – which is easily proved by changing z into $-z$ in the original array. We put this to good use.

Example 5.33

We will find a formula for the alternating sum of the first n cubes of the positive integers $\sum_{k=1}^{n}(-1)^k k^3$. We use the sum operation of Theorem 5.19 on the last result –

with a dash of partial fractions. We have

$$\left(\frac{1}{1-z}\right)\sum_{r\geq 0} r^3(-1)^r z^r = \frac{-z+4z^2-z^3}{(1-z)(1+z)^4}$$

$$= \frac{1}{8}\left(\frac{z^3+13z^2-13z-1}{(1+z)^4}+\frac{1}{1-z}\right)$$

$$= \frac{1}{8}\left((z^3+13z^2-13z-1)\sum_{r\geq 0}(-1)^r\binom{r+3}{3}z^r+\sum_{r\geq 0}z^r\right).$$

Comparing coefficients of z^n we find that

$$\sum_{k=0}^{n}(-1)^k k^3 = \frac{(-1)^n\left(\binom{n}{3}+13\binom{n+1}{3}-13\binom{n+2}{3}-\binom{n+3}{3}+(-1)^n\right)}{8}$$

$$= \frac{(-1)^n}{8}\left((-1)^n+4n^3+6n^2-1\right).$$

We end this section with an explicit form for the Eulerian numbers which may be adapted to count ascents.

Theorem 5.34 (Explicit form for Eulerian numbers)

The Eulerian numbers are given by $\left\langle{n\atop k}\right\rangle = \sum_{l=0}^{k}\binom{n+1}{k-l}(-1)^{k-1}l^n$.

Proof

We start with the result of using the Euler operation, successively, on the function $\frac{1}{1-z}$

$$\sum_{r\geq 0} r^n z^r = \frac{\sum_{k\geq 0}\left\langle{n\atop k}\right\rangle z^k}{(1-z)^{n+1}}.$$

A rearrangement gives

$$(1-z)^{n+1}\sum_{r\geq 0}r^n z^r = \sum_{k\geq 0}\left\langle{n\atop k}\right\rangle z^k$$

and using the Binomial Theorem on the left gives

$$\left(\sum_{r\geq 0}\binom{n+1}{r}(-1)^r z^r\right)\left(\sum_{r\geq 0}r^n z^r\right) = \sum_{k\geq 0}\left\langle{n\atop k}\right\rangle z^k.$$

If we compare coefficients of z^k on either side we find that

$$\sum_{l=0}^{k} \binom{n+1}{k-l} (-1)^{k-l} l^n = \binom{n}{k}$$

exactly as required. $\qquad\qquad\qquad\qquad\qquad\qquad\qquad\qquad\qquad\qquad\qquad\qquad\square$

5.4.5 Exercises

Exercise 5.13

Find a formula for the alternating sum of the first n squares of the natural numbers.

Exercise 5.14

Show that the generating function for the sequence $\{L_{r+2}\}$ is

$$\sum_{r\geq 0} L_{r+2} z^r = \frac{3+z}{1-z-z^2}.$$

Use this to show that:

(i) $\sum\limits_{k=0}^{r} L_k = L_{r+2} - 1$. (**Hint**: use the sum formula on the generating function of the Lucas numbers, together with partial fractions);

(ii) $\sum\limits_{k=0}^{r} (-3)^k L_{k+2} = 3(-3)^r F_{r+1}$. (**Hint**: divide each side of the generating function above by $3+z$ and use it as a weighted sum)

Exercise 5.15

Show that the generating function for the sequence $\{L_{r+3}\}$ is $\frac{4+3z}{1-z-z^2}$. Hence show that $\sum\limits_{k=0}^{r} 2^{r-k} L_k = 6.2^r - L_{r+3}$. (**Hint**: use the expression $1-2z$ as a weighted sum.)

Exercise 5.16

By considering the product of the generating functions $\frac{1}{1-z^2}$ and $\frac{z}{1-z-z^2}$ show that:

(i) $\sum\limits_{k=0}^{r} F_{2k} = F_{2r+1} - 1$;

(ii) $\sum\limits_{k=0}^{r} F_{2k+1} = F_{2r+2}.$

What are the corresponding Lucas results?

5.5 Miscellaneous Exercises

Exercise 5.17

By multiplying suitable generating functions find expressions for $\sum_{k=0}^{r} F_k F_{r-k}$ and $\sum_{k=0}^{r} L_k L_{r-k}$. (**Hint**: see Example 5.18.)

Exercise 5.18

Consider the sequence $\{u_r\}$ in which $u_r = rF_r$. Find a recurrence relation satisfied by the sequence. (**Hint**: apply the Euler operator to the generating function of the Fibonacci sequence.)

Exercise 5.19

Show that the generating function for the sequence $\{F_{r+2}\}$ is $\frac{1+z}{1-z-z^2}$ and use this to show that

$$F_{r+1} = \sum_{k=0}^{r} (-1)^{r-k} F_{k+2}.$$

Exercise 5.20

Prove that $2F_{r+1} = \sum_{k=0}^{r} (L_k - F_k)$ and $5F_r = L_{r+1} + L_{r-1}$.

Exercise 5.21

By applying the Euler operator to a the sum formula for a GP, prove that the formula for the "weighted" sum of a GP is given by

$$\sum_{k=0}^{n} ka^k = \frac{(n+1)a^n}{a-1} - \frac{a^{n+1}-1}{(a-1)^2} \quad (a \neq 1).$$

Exercise 5.22

This question concerns the sum $\sum_{k=1}^{r} k^2 = \frac{r(r+1)(2r+1)}{6}$.

(i) Show that $k^2 = \binom{k}{1} + 2\binom{k}{2}$ and hence use Pascal's Triangle to show that the sum takes the value given.

(ii) Now use Eulerian numbers to find the same result.

Exercise 5.23

The **Central terms** of the even indexed rows of Pascal's triangle are given by $\binom{2r}{r}$. Use an enumerative argument to prove that $\binom{2r}{r} = \sum_{k\geq 0} \binom{r}{k}^2$.

Exercise 5.24

A new array – like Pascal's triangle – whose terms are written $]_k^r[$ is defined. Its first column is the sequence $\{2, 1, 1, \ldots\}$ and subsequent terms are produced by the recurrence

$$\left]\begin{matrix}r+1\\k\end{matrix}\right[= \left]\begin{matrix}r\\k-1\end{matrix}\right[+ \left]\begin{matrix}r\\k\end{matrix}\right[$$

just like that for Pascal's triangle. What is the sequence of the diagonal sums?
Prove it.

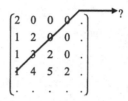

Figure 5.3 Diagonal sums of an array.

6

Group Actions and Counting

6.1 The First Steps

In this chapter we count configurations with symmetries, seeking to count only once those configurations that are essentially the same. In Chapter 4 we introduced the idea of a group action and the orbits of this action. This is going to be crucial to the development. Roughly we have an object which we are going to "colour" and we find the number of ways we can do this if we allow for the symmetries of the object.

Example 6.1

Think of the cube as a dice. We are going to count the number of ways to colour the faces with the six colours $\{1, 2, 3, 4, 5, 6\}$, insisting that each colour is used. The usual rule is that opposite sides add to seven but there are at least two ways to do this, if you do not allow reflections (see Figure 6.1). So how many ways can we colour the cube with six colours using all the colours?

Example 6.2

When we talk about permutations on a set we usually mean a labelled set. That is, we can distinguish the elements of the set. There are $r!$ permutations if the set has size r. What happens if the set was unlabelled? Then there is only one permutation, since any ordering is the same. What happens if there are four elements and just two labels,

A. Camina, B. Lewis, *An Introduction to Enumeration*,
Springer Undergraduate Mathematics Series,
DOI 10.1007/978-0-85729-600-9_6, © Springer-Verlag London Limited 2011

Figure 6.1 Two dice: different?

say d and e? It is easy to list the possibilities *ddee, dede, deed, eedd, eded* and *edde*. Labelling makes a difference: labelling and coloring are the same.

The difficulty in being precise comes from the wide range of objects we wish to count: they might be geometric, algebraic or something which might not appear to be mathematical. How does this affect the counting? The key point is that we do not want to count repeats, that is things which appear the same under the group of symmetries. The best way to proceed is to give some examples and then develop the ideas further.

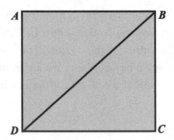

Figure 6.2 A square configuration.

Example 6.3

Let *ABCD* be a square and assume that the diagonal *BD* is joined, as in Figure 6.2.

What is the symmetry group, G, of this figure? It is obviously not the whole symmetry group of the square, which has order 8. It is not too difficult to see that there are only four symmetries which we will describe by their actions on the vertices: the identity, rotation through π, $(AC)(BD)$, reflection in the axis AC, (BD), and reflection in the axis BD, (AC). So G has order 4.

What are the orbits of this symmetry group on vertices? It is easy to see that one orbit is $\{A, C\}$ and another is $\{B, D\}$. Geometrically this is clear, as the vertices A and C have two edges joining them to other vertices, but vertices B and D have only three edges joining them to other vertices. Note that $G(A) = \{A, C\} = G(C)$ and $G(B) = \{B, D\} = G(D)$. Also we can check that $G_A = \{1, (BD)\}$ so $|G_A| = 2$ and $[G : G_A] = 2$, as required by the Orbit-Stabilizer Theorem 4.48.

We might also look at the action on edges. There are five edges. This time there are still two orbits: one consists of the outside edges and the other of the diagonal.

Theorem 6.4 (Number of orbits using fixed points)

Let G be a finite group and let G act on X. Let k be the number of orbits then

$$k = \frac{1}{|G|} \sum_{g \in G} |\mathrm{Fix}(g)|. \tag{6.1}$$

Proof

By the Orbit-Stabilizer Theorem $|\mathrm{Fix}(g)| = [G : G(g)]$. Let $x_1, x_2, \ldots, x_k$ be chosen so that each x_i is in a different orbit. Now collect the terms on the right-hand side of equation above according to which orbits they are in:

$$\frac{1}{|G|} \sum_{g \in G} |\mathrm{Fix}(g)| = \frac{1}{|G|} \sum_{1}^{k} \frac{|G|}{|G(x_i)|} \times |G(x_i)| = k$$

as required. □

We now have a corollary which we will use later.

Corollary 6.5 (Orbits on pairs)

Let G be a finite group and let G act on X. Let k be the number of orbits of G acting on $X \times X$. Then

$$k = \frac{1}{|G|} \sum_{g \in G} |\mathrm{Fix}(g)|^2. \tag{6.2}$$

Proof

The action on $X \times X$ is defined by $g(x,y) = (gx,gy)$ for all $(x,y) \in X \times X$. So the number of fixed points in this action is just $|\text{Fix}(g)|^2$ and hence the result.   □

We will discuss fixed points further in Section 7.4.

Definition 6.6 (Transitive Group)

Let G be a finite group and let G act on X. If the number of orbits is one then G is called transitive, if not it is called **intransitive**.

Corollary 6.7

Let G be a finite group and let G act on X. If G is transitive then:

$$1 = \frac{1}{|G|} \sum_{g \in G} |\text{Fix}(g)|.$$

Example 6.8

Let G be the symmetry group of the "square configuration" of Figure 6.2. Consider G acting on vertices, the group is given by $\{1, (AC)(BD), (BD), (AC)\}$ and has order 4. So Equation (6.1) says the number of orbits is given by

$$\frac{4+2+2+0}{4} = 2,$$

as we already knew.

Now consider the action of G on the edges. Then we get

$$\frac{5+1+1+1}{4} = 2,$$

again as before.

In Example 6.8 we can see a general idea: there is some object which we wish to colour and the object has a symmetry group which may act on different parts of the object. For this example it was vertices or edges. In each case the symmetry group induces a permutation group on the parts we want to colour. So in our generalization we assume that the set we wish to colour comes equipped with a permutation group. However, we do not specify where the permutation group comes from. So in any particular application we have to begin by determining the group.

We begin with a counting result at the heart of this development which is a consequence of the Orbit-Stabilizer Theorem.

Theorem 6.9 (Number of orbits using stabilizers)

Let G be a finite group and let G act on X. Let k be the number of orbits then

$$k = \frac{1}{|G|} \sum_{x \in G} |G_x|. \tag{6.3}$$

Proof

If x, y are in the same orbit then $|G_x| = |G_y|$ by the Orbit-Stabilizer Theorem. So each orbit contributes $|G_x| \times |G(x)|$. But this is just $|G|$. The result now follows. $\square$

6.1.1 Exercises

Exercise 6.1

Let G be a transitive permutation group of degree greater than 1. Prove that there is an element in G with no fixed points.

Exercise 6.2

Let G act on X. Further assume that only the identity has fixed points. Prove that $|G|$ divides $|X|$ and that G is transitive if and only if $|G| = |X|$.

Exercise 6.3

Let $\mathcal{H}$ be a regular hexagon with vertices A, B, C, D, E, F labelled clockwise. Join the pairs A, E and B, D. Calculate the symmetry group G of $\mathcal{H}$ and count the number of orbits on vertices and edges.

Exercise 6.4

Let G be the symmetry group of the square. Let G act on X where X is the set of pairs of vertices. Find the number of orbits and their sizes.

Exercise 6.5

Let the cyclic group of order p generated by g act on the set of all p-tuples of elements from $\{1, 2, \ldots, n\}$ by defining

$$g(x_1, x_2, \ldots, x_p) = (x_p, x_1, \ldots, x_{p-1}).$$

By counting orbits, prove that $n^p \equiv n \pmod{p}$.

6.2 Colourings and the Cycle Index

Now we come to an important application of the ideas we have developed – given an object, in how many ways can it be coloured? Clearly the answer will depend on the number of colours and the symmetry of the object. We start with a simple version of this problem.

Example 6.10

Given a rod and some beads of different colours. How many ways can we construct a rod with two beads on it where we do not mind whether we start from the left or right? Either we have two beads of the same colour or two of different colours. Also the number of ways depends on the number of colours. If the number of colours is k there are k ways with one colour and $\binom{k}{2}$ with two colours. Note that the group is very simple, just of order 2.

Example 6.11

Now for a slightly more complicated case. Take a triangle with three beads, one at each vertex. The symmetry group of a triangle, which is just a 3-gon, has order 6.

Either we have three beads the same colour, two the same or all different. Notice because of the symmetry it doesn't matter which one is different when only two colours are used. So the number of ways depends on the number of colours. Let the number of colours be k. Then there are k ways with one colour. It is possible to choose two colours in $\binom{k}{2}$ ways. But choosing two red and one black, say, is different from choosing two black and one red. So there are $k(k-1)$ ways to colour the triangle with two colours. Finally, with three colours there are just $\binom{k}{3}$ ways.

Note that in these two examples, if we had not been concerned with the symmetries the numbers would have been k^2 and k^3 respectively.

To develop the ideas of colouring we need some ingredients and a recipe.

Definition 6.12 (Colouring)

Given a set X with a group G acting on it, a set of colours (say C) and finally a map from $C \to X$, which is the colouring. The set of all maps, colourings, will be denoted by $\mathcal{C}$. We will assume that $X = n$ and $|C| = k$.

Lemma 6.13

In the situation given by Definition 6.12, G acts on the set of all colourings $\mathcal{C}$ in the following way:

$$\text{if } c \in C \text{ and } g \in G \text{ then } g(c)(x) = c(gx) \text{ for all } x \in X.$$

Then the number of colourings is the number of orbits of G on C.

Proof

To see that this defines an action, we observe that

$$g(h(c(x))) = g(c(h(x))) = c(gh(x)),$$

for all $g \in G$, $c \in C$ and $x \in X$.

To say that two colorings are equivalent is saying that the group action maps one to the other. □

Definition 6.14 (Cycle index)

Let G be a permutation group of degree n. Let $X_1, X_2, \ldots, X_n$ be n indeterminates. Let $g \in G$ and let g have n_k cycles of length k. Then the term $\prod_1^n X_k^{n_k}$ is called the cycle symbol, labelled $cs(g)$ of g. Then the cycle index of G is

$$P_G[X_1, X_2, \ldots, X_n] = \frac{\sum_{g \in G} cs(g)}{|G|}.$$

We often drop the indeterminates and just write P_G.

The cycle index can be thought of as a type of generating function. It is interesting to note that we always divide through by the order of G. So all these expressions are taking an "average".

Example 6.15

Start with the simplest example, $G = \text{Sym}(2)$. Note that the identity element gives X_1^n where n is the degree. The only other element is the transposition (12), which gives X_2, one cycle of length two. So

$$P_G = \frac{X_1^2 + X_2}{2}.$$

Example 6.16

This time let G be $\text{Sym}(3)$. So G has six elements and its degree is 3. The identity element gives X_1^3, this time a transposition gives $X_1 X_2$ and the three cycles give X_3. Since

G has 3 transpositions and 2 three cycles (we get this count from Proposition 2.3),

$$P_G = \frac{X_1^3 + 3X_1X_2 + 2X_3}{6}.$$

The importance of this idea is evident when we think about the Orbit-Stabilizer Theorem. In this we need to know how many things an element fixes. So what happens when we fix an element $g \in G$ and try to see which colourings it fixes? If g fixes a colouring and two elements of X are in the same cycle of g they must share the same colour. Since each cycle is independent of every other cycle we can colour each cycle independently. If there are m cycles altogether and k colours there are k^m ways this can be done. This is the value of $cs(g)$ with each indeterminate replaced by k.

Proposition 6.17

Let G act on X and let C be a the set of colourings from a set of colours C, $|C| = k$. Let c_m be the number of elements of G with exactly m cycles. Then the number of orbits of G on C is given by:

$$\frac{\sum_1^\infty c_m k^m}{|G|}.$$

Proof

This now follows immediately from Theorem 6.4.   □

Example 6.18

In Example 6.15 we can put $X_i = k$, $i = 1, 2$ and get

$$P(k,k) = \frac{k^2 + k}{2}.$$

In Example 6.10 we counted k colourings with one colour used and $\binom{k}{2}$ colourings when both colours were used. We note that

$$k + \binom{k}{2} = \frac{k^2 + k}{2}.$$

Example 6.19

In Example 6.16 we can put $X_i = k$ for each i and get

$$P(k,k) = \frac{k^3 + 3k^2 + 2k}{6}.$$

In Example 6.11 we counted the colourings with k colours. After a little algebra it can be seen that the numbers agree.

Example 6.20

Consider Figure 6.3, in which we have six different representations of bracelets made from four beads of two different colours. Can we prove that these are the only possibilities?

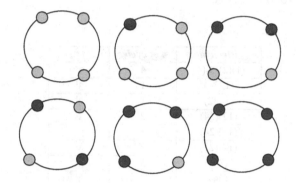

Figure 6.3 Bracelets with four beads and two colours.

The symmetry group is the symmetry group of the square, which is of order 8. If we assume that the beads are labelled with 1, 2, 3 and 4 we can write out the elements as in Table 6.1. Note that for the last two entries, although only one cycle is shown,

Element	# of cycles	k^m
(1)(2)(3)(4)	4	2^4
(1234)	1	2^1
(1432)	1	2^1
(12)(34)	2	2^2
(13)(24)	2	2^2
(14)(23)	2	2^2
(13)	3	2^3
(24)	3	2^3

Table 6.1 Table for bracelets with four beads and two colours.

there are two fixed points which give 2 one cycles. Now

$$P_G(2,2,2,2) = \frac{2^4 + 2.2^1 + 3.2^2 + 2.2^3}{8} = 6.$$

We get 6 as we anticipated, confirming that these are the only possibilities.

Example 6.21

If we consider the previous example it is easy to see what happens if there are three colours as in the following table:

Element	# of cycles	k^m
(1)(2)(3)(4)	4	3^4
(1234)	1	3^1
(1432)	1	3^1
(12)(34)	2	3^2
(13)(24)	2	3^2
(14)(23)	2	3^2
(13)	3	3^3
(24)	3	3^3

Table 6.2 Table for bracelets with three colours.

Now

$$P_G(3,3,3,3) = \frac{3^4 + 2.3^1 + 3.3^2 + 2.3^3}{8} = 21.$$

All these examples show that this only goes so far – it calculates the total number of colourings but does not detail how many there are with further restrictions. For example we might want to know how many have exactly two white beads and how many with two black beads. This is our next task.

6.2.1 Exercises

Exercise 6.6

In how many ways can we paint the faces of a cube using two colours if we allow only rotations of the cube?

Exercise 6.7

Calculate the cycle index of the group Alt(4).

Exercise 6.8

Consider the situation in Exercise 6.3. Calculate the cycle index of the symmetry group. Then calculate how many colourings, with k colours, there are of the vertices.

Exercise 6.9

In Exercise 4.14 we calculated the symmetry group of the tetrahedron. Now calculate the cycle index of the symmetry group. Show that with two and three colours, respectively, there are 5 and 15 colourings of the vertices and faces and 11 and 66 colourings of the edges.

Exercise 6.10

Consider the square and the action of the symmetry group on pairs of vertices, as in Exercise 6.4. Find the cycle index for this action.

6.3 Polya's Theorem

We begin this section by defining an expression that counts the number of colourings with different configurations.

Definition 6.22 (Colouring inventory)

Given a colouring with $C = \{c_1, c_2, \ldots, c_k\}$ the set of colours and G acting on X. Let $\mathcal{C}$ be the set of all possible colourings.

The colouring inventory of a subset $\mathcal{D} \subset \mathcal{C}$ is a polynomial in $\{c_1, c_2, \ldots, c_k\}$ which is the sum of terms of the form $\prod c_i^{d_i}$ with coefficients which count how many colourings there are in $\mathcal{D}$ with d_i elements of X coloured c_i.

Example 6.23

Let us reconsider Example 6.20. First consider all the possible bracelets, and use $C = \{b, w\}$. We will calculate the colouring inventory polynomial by exhaustion, that is we will list all the possibilities. Those which have exactly one colour will be the terms b^4 and w^4; with two colours, one occurring three times gives $4wb^3$ and $4bw^3$, since there are four choices to allocate the single colour; finally, there are six possibilities for two colours each occurring twice, since we can choose any two positions for the two colours, giving $6b^2w^2$. Notice in this count we do not look at the action of the group. So the colouring inventory polynomial is

$$b^4 + 4b^3w + 6b^2w^2 + 4bw^3 + w^4 = (b+w)^4.$$

We could look at the subset fixed by a permutation, say $g = (1,3)$. This will fix those colourings which have a particular diagonal pair the same colour, assuming the labelling was clockwise. So the corresponding colouring inventory polynomial will be

$$b^4 + 3b^3w + 4b^2w^2 + 3bw^3 + w^4.$$

Definition 6.24 (Pattern inventory)

Given a colouring with G acting on X, with $C = \{c_1, c_2, \ldots, c_k\}$, the set of colours. Let C be the set of all possible colourings. The pattern inventory of a subset $\mathcal{D} \subset C$ is a polynomial in $\{c_1, c_2, \ldots, c_k\}$ which is the sum of terms of the form $\prod c_i^{d_i}$ with coefficients which count how many **equivalent** colourings there are in $\mathcal{D}$ with d_i elements of X coloured c_i.

The important point here is that the pattern inventory takes account of the group action.

Example 6.25

Let us reconsider, yet again, Example 6.20. Consider all the possible bracelets, using again the principle of exhaustion, with $C = \{b, w\}$. The pattern inventory will be given by

$$b^4 + b^3w + 2b^2w^2 + bw^3 + w^4.$$

Now the question is "How do we find this without enumerating all possibilities?".

Recall that by Theorem 6.4 the number of orbits depends on the size of the fixed sets of the elements of the group. Here is a key lemma.

Lemma 6.26

Let $C = \{c_1, c_2, \ldots, c_k\}$ be a set of colours with G acting on X. Let $g \in G$ then the colouring inventory for the set $\text{Fix}(g)$ is given by finding the cycle index for g, $cs(g)$ and substituting $\sum c_j^r$ for X_r.

Proof

Consider one cycle, say $(a_1 a_2 \ldots a_r)$ of g. Then if g fixes a colouring then each of $a_1, a_2, \ldots, a_r$ has to have the same colour. So the pattern inventory for this cycle will be $c_1^r + c_2^r + \cdots + c_k^r$. In $cs(g)$ this cycle contributes the term X_r. For each cycle we can do the same, and as each cycle is disjoint, the choices for each cycle can be multiplied together. So the total number of colourings fixed by g is the product

$$\prod_{\text{each cycle}} c_1^r + c_2^r + \cdots + c_k^r.$$

This is the same as the value of $cs(g)$ with each indeterminate X_r replaced by $c_1^r + c_2^r + \cdots + c_k^r$. The lemma follows. $\qquad\square$

Theorem 6.27 (Polya's Theorem)

Let $C = \{c_1, c_2, \ldots, c_k\}$ be a set of colours with G acting on X. Then the pattern inventory is given by substituting $\sum c_j^r$ for X_r. Thus the pattern inventory is

$$P_G\left(\sum c_j, \sum c_j^2, \ldots, \sum c_j^k\right).$$

Proof

We use the result of Theorem 6.4. For any particular colouring, to find the number of equivalence classes, we add the numbers of fixed points and divide by the order of G. But by the previous lemma the inventory for Fix$g$ is just $cs(g)$ with the appropriate substitution and the average is just the definition of P_G. $\qquad\square$

Example 6.28

Once more we reconsider Example 6.20. First we need to calculate P_G, which is quite straightforward:

$$P_G = \frac{1}{8}\left(X_1^4 + 2X_1^2X_2 + 3X_2^2 + 2X_4\right).$$

For two colours $\{b, w\}$ the pattern inventory is

$$\frac{1}{8}\left((b+w)^4 + 2(b+w)^2(b^2+w^2) + 3(b^2+w^2)^2 + 2(b^4+w^4)\right).$$

After some expansion we get $b^4 + b^3w + 2b^2w^2 + bw^3 + w^4$, which is precisely the same result we established earlier by direct enumeration.

Example 6.29

We will count the number of ways we can colour a square board with black and white squares. We can count this using Polya's Theorem 6.27. We need to calculate the cyclic index polynomial for the symmetry group acting on the board. This is a group of order 8, the symmetry group of the square. In this example we will only consider the case when the board has an even number of sides $2n \times 2n$. The only automorphism which fixes any squares are the reflections in the diagonal. So we get

$$P = \frac{1}{8}\left(Z_1^{4n^2} + 2Z_4^{n^2} + 3Z_2^{2n^2} + 2Z_1^{2n}Z_2^{\frac{4n^2-2n}{2}}\right).$$

So replacing the indeterminates by 2 we get

$$\frac{1}{8}\left(2^{4n^2} + 2.2^{n^2} + 3.2^{2n^2} + 2.2^{2n^2+n}\right).$$

6.3.1 Exercises

Exercise 6.11

Sym(3) is the symmetry group of the triangle. Calculate the pattern inventory for colouring the vertices with two colours, say red and green. How many have two red vertices?

Exercise 6.12

We colour a necklace with four beads using three distinct colours. How many will use all three colours?

Exercise 6.13

Calculate the cycle index of Sym(4) acting on

(i) unordered pairs of distinct elements of $\{1,2,3,4\}$
   and

(ii) ordered pairs of elements of $\{1,2,3,4\}$.

Exercise 6.14

We now colour a necklace with five beads using two colours, say R and G. Show that the colourings are given by the following polynomial:

$$R^5 + R^4G + 2R^3G^2 + 2R^2G^3 + RG^4 + G^5.$$

6.4 Miscellaneous Exercises

Exercise 6.15

Let G act on a set X and let p be a prime. Assume that for each $x \in X$ there is a p-subgroup P_x which fixes precisely the element x. Prove that G is transitive.

Exercise 6.16

Let C be a cyclic group of order p, for some prime p. Consider the regular action of C on itself. Prove:

$$P_C(X_1, X_2, \ldots, X_p) = \frac{1}{p} \left(X_1^p + (p-1)X_p \right).$$

Hence show that $k^p \equiv k \pmod{p}$ for any $k \in \mathbb{N}$.

Exercise 6.17

Let G be the group of rotational symmetries of the cube. Calculate the cycle index for G acting on the faces. Then calculate the pattern inventory for colouring the faces with red, white and blue. Show that there are six different ways to use both colours twice.

Exercise 6.18

This final exercise is an application to music. In twelve tone rows the musician is allowed to choose a row of distinct notes from the twelve tone scale. Two rows are considered the same if they can be transformed by transposing or inverting. The question is, given a row of say six notes, how many rows are we allowed? This can be translated as follows. The possible notes, called pitch classes, are labelled from 0 to 11, so we think of them as the integers mod 12. The twelve transformations are called $T_n : x \to x+n \pmod{12}$ and $T_n I : x \to -x+n \pmod{12}$. Show that the numbers in the table below are correct.

Length of row	0	1	2	3	4	5	6	7	8	9	10	11	12
Number of classes	1	1	6	12	29	38	50	38	29	12	6	1	1

Exponential Generating Functions

7.1 Another Generating Function

It is not clear why another generating function is needed, but it turns out that when we count the number of permutations of different types, these are the natural generating function. We can see this by looking at a simple enumeration: the number of permutations of r things.

Example 7.1

We will attempt to find the generating function of the sequence $\{u_r\} = \{r!\}$. Written as a power series, the generating function is

$$F(z) = 1 + 1z + 2z^2 + 6z^3 + 24z^4 + \cdots.$$

And that is as far as we can go. There is nothing simple we can do with this to find a function whose expansion this is. But that isn't the only problem. The ratio test tells us that such a function has a zero radius of convergence: it converges only when $z = 0$. Despite this, we can still extract enumerative information from it, but not with the benefit of an analytic function in some of the ways we have done hitherto.

Definition 7.2 (Exponential generating function *egf*)

Given any sequence $\{u_r\} = \{u_0, u_1, u_2, \ldots\}$ the exponential generating function (*egf*)

A. Camina, B. Lewis, *An Introduction to Enumeration*,
Springer Undergraduate Mathematics Series,
DOI 10.1007/978-0-85729-600-9_7, © Springer-Verlag London Limited 2011

of the sequence is the function $U(z)$ in which

$$U(z) = \frac{u_0}{0!} + \frac{u_1 z}{1!} + \frac{u_2 z^2}{2!} + \cdots = \sum_{r \geq 0} \frac{u_r z^r}{r!}.$$

Note: we refer to the generating function

$$U(z) = \sum_{r \geq 0} u_r z^r$$

as the ordinary generating function (*ogf*) of the sequence. This definition solves the problem of finding a generating function for permutations.

Example 7.3

The exponential generating function (*egf*) of the sequence, $\{u_r\} = \{r!\}$ is given by

$$U_{egf}(z) = \sum_{r \geq 0} \frac{u_r z^r}{r!} = \sum_{r \geq 0} \frac{r! z^r}{r!} = \sum_{r \geq 0} z^r = \frac{1}{1-z},$$

which has radius of convergence $|z| < 1$.

Example 7.4

It is useful to see the difference between an ordinary generating function (*ogf*) and the corresponding exponential generating function (*egf*) of a particular sequence; we take the simple sequence $\{u_r\} = \{1, 1, 1, \ldots\}$. We have:

(i) the *ogf* is $U_{ogf}(z) = \sum_{r \geq 0} u_r z^r = \sum_{r \geq 0} z^r = \frac{1}{1-z}$;

(ii) the *egf* is $U_{egf}(z) = \sum_{r \geq 0} \frac{u_r z^r}{r!} = \sum_{r \geq 0} \frac{z^r}{r!} = e^z$.

There is an important enumerative difference between the two types of generating function. Here is a cautionary but illuminating example that highlights the difference.

Example 7.5

We find the number of ways that three letters, chosen from the letters of SEQUENCE, can be arranged. The ordinary generating function for the choice of the three possible Es is $1 + z + z^2 + z^3$; that for each of the letters C, N, S, Q and U is $1 + z$. So the generating function is

$$(1+z)^5 (1 + z + z^2 + z^3)$$

and the answer we seek is the coefficient of z^3, which is

$$1 + 5.1 + 10.1 + 10.1 = 26.$$

But what we have calculated is the number of non-negative integer solutions of the equation $s + e + q + u + n + c = 3$, in which $0 \leq e \leq 3$ and $0 \leq s, q, u, n, c \leq 1$. This is **not** the answer. A selection of three letters from SEQUENCE has (by the principle of exhaustion) either none, one or two Es (Table 7.1) and so:

Construction		Number of 'word types'	Number of 'words'	Example
3 Es:	EEE	1	1	EEE
2Es +1:	EES EEN EEU EEQ EEC	5	$5 \cdot \frac{3!}{2!} = 15$	EES, ESE, SEE, ...
1E + 2:	ESQ EQU EUN ENC ESU EQN EUC ESN EQC ESC	10	$10 \cdot 3! = 60$	ESQ,EQS, SEQ, SQE, QES, QSE,...
0 Es + 3:	SQU SUN SNC QUN QNC UNC SQN SUC QUC SQC	10	$10 \cdot 3! = 60$	SQU, SUQ, QSU, QUS, USQ, UQS,...
	Total	26	136	

Table 7.1 Arranging the letters of SEQUENCE.

The distinction that needs to be made is between:

 (i) making a selection;

(ii) making a selection and then arranging (or if you prefer, labelling) those selected.

We obtain the count of the number of "words" from an *egf*. The exponential generating function for the letter E is

$$1 + \frac{z}{1!} + \frac{z^2}{2!} + \frac{z^3}{3!};$$

that for each of the letters C, N, S, Q and U is

$$1 + \frac{z}{1!} = 1 + z.$$

So the required *egf* is

$$(1 + z)^5 \left(1 + \frac{z}{1!} + \frac{z^2}{2!} + \frac{z^3}{3!}\right).$$

We seek the coefficient of $\frac{z^3}{3!}$ and this is $1 + 5.1.3 + 10.1.6 + 10.6 = 136$.

We explore some some particular sequences and their *egf*s.

Example 7.6

We find the *egf* of the sequence $\{u_r\} = \{r^2\}$. We may write $r^2 = r(r-1) + r$ and then

$$\sum_{r\geq 0} \frac{u_r z^r}{r!} = \sum_{r\geq 0} \frac{r^2 z^r}{r!}$$

$$= \sum_{r\geq 0} \frac{(r(r-1)+r) z^r}{r!} = \sum_{r\geq 2} \frac{z^r}{(r-2)!} + \sum_{r\geq 1} \frac{z^r}{(r-1)!}$$

$$= z^2 \sum_{r\geq 2} \frac{z^{r-2}}{(r-2)!} + z \sum_{r\geq 1} \frac{z^{r-1}}{(r-1)!} = z^2 e^z + z e^z$$

$$= (z^2 + z) e^z.$$

Theorem 7.7 (*egf* of Fibonacci and Lucas sequences)

The *egf* for the Fibonacci and Lucas sequences are

(i) $\sum_{r\geq 0} \frac{F_r z^r}{r!} = \frac{1}{\sqrt{5}} \left(e^{\phi_1 z} - e^{\phi_2 z} \right);$

(ii) $\sum_{r\geq 0} \frac{L_r z^r}{r!} = e^{\phi_1 z} + e^{\phi_2 z}.$

Proof

We start with Binet's formula

$$\sum_{r\geq 0} \frac{F_r z^r}{r!} = \sum_{r\geq 0} \frac{1}{\sqrt{5}} \frac{(\phi_1^r - \phi_2^r) z^r}{r!}$$

$$= \frac{1}{\sqrt{5}} \left(\sum_{r\geq 0} \frac{(\phi_1 z)^r}{r!} - \sum_{r\geq 0} \frac{(\phi_2 z)^r}{r!} \right)$$

$$= \frac{1}{\sqrt{5}} \left(e^{\phi_1 z} - e^{\phi_2 z} \right).$$

The proof for the Lucas numbers is in the exercises. □

7.1.1 Exercises

Exercise 7.1

In how many ways can four letters chosen from the letters of FUNCTION be arranged?

Exercise 7.2

Find the *egf* of the sequence whose terms consist of:

(i) the Lucas numbers L_r (this completes the proof of Theorem 7.7);

(ii) the even indexed Fibonacci numbers, F_{2r}.

Exercise 7.3

Consider the sequence $\{u_r\}$ in which $u_r = r^3$. By writing the terms of the sequence in the form

$$u_r = r^3 = Ar(r-1)(r-2) + Br(r-1) + Cr$$

for suitable constants, A, B and C, find the *egf* of the sequence.

7.2 Recurrence to *egf*

The basic procedure for converting a recurrence into an exponential generating function does not differ from the corresponding procedure for an ordinary generating function – the three-step recipe (Algorithm 2.23). Of course, we need to introduce the factorial term as part of the summation. We illustrate it with a sequence we have encountered on more than one occasion: the derangement sequence.

Example 7.8 (The derangement sequence)

We will use two recurrences that we have for this sequence: Example 2.19 and its equivalent from Exercise 2.8.

(i) The first step is to write out the recurrence

$$d_{r+1} = r(d_r + d_{r-1}) \text{ (noting that } d_0 = 1, d_1 = 0.)$$

(ii) The next step is quite straightforward (remember the new factorial element):

$$\sum_{r\geq 1} \frac{d_{r+1}z^r}{r!} = \sum_{r\geq 1} \frac{rd_r z^r}{r!} + \sum_{r\geq 1} \frac{rd_{r-1}z^r}{r!}.$$

(iii) Suppose the *egf* required is $D(z)$ then

$$D'(z) - d_1 = z\sum_{r\geq 1} \frac{d_r z^{r-1}}{(r-1)!} + z\sum_{r\geq 1} \frac{d_{r-1}z^{r-1}}{(r-1)!}$$
$$D'(z) = zD'(z) + zD(z).$$

What we've ended up with is a differential equation! Luckily, it is one in which we can separate the variables:

$$\frac{D'(z)}{D(z)} = \frac{z}{1-z} = \frac{1}{1-z} - 1;$$

we integrate, and find that

$$\log(D(z)) = -\log(1-z) - z + C.$$

The constant of integration is zero since $D(0) = d_0 = 1$, and then we have

$$D(z) = \frac{e^{-z}}{1-z}.$$

Note: differential equations frequently appear in the final part of the three-step recipe for an *egf*. They are not always easily solved.

Here is another way of deriving the *egf* above.

(i) We start by writing out the other recurrence:

$$d_r = rd_{r-1} + (-1)^r \text{ with } d_0 = 1;$$

(ii) The second step then becomes

$$\sum_{r\geq 1} \frac{d_r z^r}{r!} = \sum_{r\geq 1} \frac{rd_{r-1}z^r}{r!} + \sum_{r\geq 1} \frac{(-1)^r z^r}{r!};$$

(iii) Suppose the *egf* required is $D(z)$ – then the decisive step is

$$D(z) - d_0 = zD(z) + e^{-z} - 1$$

and so

$$(1-z)D(z) = e^{-z}.$$

Finally we "solve" for $D(z)$:

$$D(z) = \frac{e^{-z}}{1-z}.$$

7.2.1 Exercises

Exercise 7.4

Attempt to convert the recurrence $t_{r+2} = t_{r+1} + (r+1)t_r$ with $t_0 = t_1 = 1$ into an *egf*. Now write it in the form $t_{r+1} = t_r + rt_{r-1}$ and repeat the process. (**Moral**: you sometimes have to be creative.)

Exercise 7.5

The push-button lock sequence $\{c_r\}$ obeys the recurrence $c_r = rc_{r-1} + 1$ (Exercise 5.6). Use this to find its *egf*.

7.3 Operations on *egf*s

We can do all the things with an *egf* that we can do with an *ogf*; we often do them in different ways though. We start with the operations that we can perform on *egf*s. This again helps with the final step in the three-step recipe of converting a recurrence into a generating function – now adapted to exponential generating functions. It provides a repertoire of techniques that we can exploit, just as before. Many of the operations we have already encountered for *ogf*s apply to *egf*s and these are not repeated. Others are adapted to a new purpose. Some have restricted use in this new environment: the sum operator is one such example.

7.3.1 Differentiation

If we differentiate the exponential generating function

$$U(z) = \sum_{r \geq 0} \frac{u_r}{z^r} r!,$$

we find that

$$U'(z) = \sum_{r \geq 1} \frac{ru_r z^{r-1}}{r!} = \sum_{r \geq 1} \frac{u_r z^{r-1}}{(r-1)!};$$

we may re-index this sum $r-1 \to r$ so that,

$$U'(z) = D(U(z)) = \sum_{r \geq 0} \frac{u_{r+1} z^r}{r!}.$$

This proves an important result.

Theorem 7.9 (Derivative of an *egf*)

The derivative of the *egf* of a sequence $\{u_r\}$ is the *egf* for the sequence $\{u_{r+1}\}$.

This operation is frequently used to derive recurrence relations.

Example 7.10

By differentiating the *egf* of the push-button lock sequence $\{c_r\}$, which has the *egf* $\frac{e^z}{1-z}$ (and using the previous theorem), we find that

$$\sum_{r\geq 0}\frac{c_{r+1}z^r}{r!} = \frac{(1-z)e^z + e^z}{(1-z)^2} = \frac{e^z}{1-z} + \frac{1}{1-z}\left(\frac{e^z}{1-z}\right)$$

$$= \sum_{r\geq 0}\frac{c_r z^r}{r!} + \frac{1}{1-z}\sum_{r\geq 0}\frac{c_r z^r}{r!}$$

and then we have

$$(1-z)\sum_{r\geq 0}\frac{c_{r+1}z^r}{r!} = (1-z)\sum_{r\geq 0}\frac{c_r z^r}{r!} + \sum_{r\geq 0}\frac{c_r z^r}{r!}.$$

If we compare coefficients of z^r on either side this gives

$$\frac{c_{r+1}}{r!} - \frac{c_r}{(r-1)!} = \frac{c_r}{r!} - \frac{c_{r-1}}{(r-1)!} + \frac{c_r}{r!}$$

and then multiplication throughout by $r!$ gives

$$c_{r+1} - rc_r = c_r - rc_{r-1} + c_r$$

and finally, the recurrence

$$c_{r+1} = (r+2)c_r - rc_{r-1}.$$

Notice that this recurrence is not the one you used to derive the *egf* for this sequence in Exercise 7.5, but we can easily derive it from the latter recurrence.

Example 7.11

We start with the recurrence

$$c_r = rc_{r-1} + 1$$

and hence (increasing the index by 1)

$$c_{r+1} = (r+1)c_r + 1;$$

if these are subtracted and simplified, we find that

$$c_{r+1} = (r+2)c_r - rc_{r-1}.$$

7.3.2 Products of Exponential Generating Functions

The result for the product of two *egfs* (or the square of one) has an unexpected new element.

Theorem 7.12 (Product formula for two *egfs*)

In the product of two exponential generating functions $W(z) = U(z)V(z)$, which are the *egfs* of the respective sequences $\{w_r\}$, $\{u_r\}$ and $\{v_r\}$, we have

$$w_r = \sum_{k \geq 0} \binom{r}{k} u_k v_{r-k}.$$

Proof

We have:

$$U(z)V(z) = \left(\sum_{r \geq 0} \frac{u_r z^r}{r!} \right) \left(\sum_{r \geq 0} \frac{v_r z^r}{r!} \right).$$

This product is itself a generating function $W(z)$ for the sequence $\{w_r\}$ in which:

$$\sum_{r \geq 0} \frac{w_r z^r}{r!} = \left(\sum_{r \geq 0} \frac{u_r z^r}{r!} \right) \left(\sum_{r \geq 0} \frac{v_r z^r}{r!} \right)$$

and comparing coefficients of z^r on either side gives

$$\frac{w_r}{r!} = \sum_{k=0}^{r} \frac{u_k v_{r-k}}{k!(r-k)!}$$

which on multiplication throughout by $r!$ gives the required result. $\qquad\square$

This product formula, and its predecessor, the product formula for *ogfs*, provide the means to describe the enumerative difference between the two generating functions:

(i) an *ogf* counts objects of size r which are are obtained by bringing together objects of size k and $r - k$ in all possible ways;

(ii) an *egf* counts objects of size r where we not only bring together two objects in the way above, but we also **relabel** the objects. There are $\binom{r}{k}$ ways to choose the new labels of the elements of the object a_k (or indeed b_{r-k} because of the symmetry of Binomial coefficients).

The product formula is a very practical result and it is often used to derive *egfs* without recourse to the three-step recipe.

Example 7.13

We know (from Equation 1.2) that the Derangement numbers satisfy the recurrence:

$$r! = \sum_{k=0}^{r} \binom{r}{k} d_k.$$

This is instantly recognisable as the result of a product of *egf*s, the product of the *egf*s of the two sequences $\{1\}$ and $\{d_r\}$ on the right, which must equal the *egf* of the sequence $\{r!\}$ on the left. So we must have

$$\frac{1}{1-z} = e^z D(z)$$

where $D(z)$ is the *egf* of the Derangement sequence. It now follows that

$$D(z) = \frac{e^{-z}}{1-z}.$$

We frequently use this idea later in the chapter. One more example, with a twist.

Example 7.14

In Exercise 5.12, we encountered the **Bell sequence** $\{B_r\}$ which was defined by the recurrence it satisfied:

$$B_{r+1} = \sum_{k=0}^{r} \binom{r}{k} B_k.$$

But this is the result of another product, except that this time, the *egf* on the left is a derivative. So

$$B'(z) = e^z B(z).$$

This is a differential equation and since $B_0 = 1$ its (unique) solution is

$$B(z) = \exp(e^z - 1)$$

and so that must be the *egf* of the sequence.

This sequence is called the Bell sequence after *E. T. Bell*. It counts the number of ways in which a set can be partitioned into non-empty, pairwise disjoint subsets. There is an interesting enumerative argument for the recurrence.

Example 7.15

Suppose we have a set with r elements which we can partition in u_r ways. If we introduce a new element then it can be partitioned in u_{r+1} ways. But we can also partition the enlarged set of elements by attaching the new element to each of the

u_r partitioned subsets. By the principle of exhaustion, the new element is *either* in a subset by itself *or* in a subset with one other element *or* with a subset with two other elements, and so on. If the subset to which the new element is adjoined has k elements then there are $\binom{r}{k} u_{r-k}$ ways to do this (select k, partition the remainder). Hence

$$u_{r+1} = \sum_{k \geq 0} \binom{r}{k} u_{r-k} = \sum_{k \geq 0} \binom{r}{k} u_k$$

since the Binomial coefficients are symmetric.

The number of partitions of a set with 1 element is 1, and hence $B_r = u_r$ because they share the same recurrence and initial values. So the Bell numbers enumerate the non-empty, pairwise disjoint partitions of a set into subsets. From Exercise 5.12 we know that

$$\{B_r\} = \{1, 1, 2, 5, 15, 52, \ldots\}.$$

Next we look at products of *egfs* as a way to derive identities.

Example 7.16

We consider the product of e^z and $L(z)$ (the *egf* of the Lucas sequence) and use it to derive an identity. We look at the product in two different ways:

(i) $e^z L(z) = e^z \left(e^{\phi_1 z} + e^{\phi_2 z} \right) = e^{(\phi_1 + 1)z} + e^{(\phi_2 + 1)z}$
and since $\phi_i + 1 = \phi_i^2$, $i = 1, 2$ (Exercise 3.16) we then have

$$e^z L(z) = e^{\phi_1^2 z} + e^{\phi_2^2 z} = \sum_{r \geq 0} \frac{\left(\phi_1^{2r} + \phi_2^{2r} \right) z^r}{r!} = \sum_{r \geq 0} \frac{L_{2r} z^r}{r!};$$

(ii) but we also have

$$e^z L(z) = \left(\sum_{r \geq 0} \frac{z^r}{r!} \right) \left(\sum_{r \geq 0} \frac{L_r z^r}{r!} \right).$$

Using the product theorem (Theorem 7.12) now gives the required identity

$$L_{2r} = \sum_{k \geq 0} \binom{r}{k} L_k.$$

7.3.3 Expanding *egfs*

An *egf* can be expanded just as their *ogf* counterparts are expanded – exactly the same techniques are employed. When we do this for the Bell sequence we find a surprising sum for each of its terms.

Example 7.17

The Bell sequence $\{B_r\}$ has *egf*

$$\sum_{r \geq 0} \frac{B_r z^r}{r!} = \exp(e^z - 1).$$

We may rearrange this to get

$$\sum_{r \geq 0} \frac{B_r z^r}{r!} = \exp(e^z - 1) = e^{-1} \exp(e^z) = e^{-1} \sum_{s \geq 0} \frac{e^{zs}}{s!} = e^{-1} \sum_{s \geq 0} \frac{1}{s!} \sum_{r \geq 0} \frac{z^r s^r}{r!}$$

and hence

$$\sum_{r \geq 0} \frac{B_r z^r}{r!} = e^{-1} \sum_{r \geq 0} \left(\sum_{s \geq 0} \frac{s^r}{s!} \right) \frac{z^r}{r!}.$$

The reversal of the infinite sums is justified since each of the sums is absolutely convergent. If we now compare coefficients of z^r on either side we have,

$$B_r = e^{-1} \sum_{s \geq 0} \frac{s^r}{s!} = e^{-1} \left(\frac{1^r}{1!} + \frac{2^r}{2!} + \frac{3^r}{3!} + \cdots \right).$$

This is the explicit form for the terms of the sequence – in the form of a convergent infinite sum. This sum is called **Dobinsky's formula**.

7.3.4 Exercises

Exercise 7.6

The derangement sequence $\{d_r\}$ has the *egf* $\frac{e^{-z}}{1-z}$. By regarding the right-hand side as the product of two *egf*s and using the product formula (Theorem 7.12), find an explicit form for the derangement numbers. (We derived this enumeratively in Example 1.27.)

Exercise 7.7

The sequence $\{t_r\}$ has the *egf* $\sum_{r \geq 0} \frac{t_r z^r}{r!} = e^{z + \frac{z^2}{2}}$. Show that

$$\sum_{k=0}^{2r+1} (-1)^{k+1} \binom{2r+1}{k} t_k = 0$$

and find the corresponding result for even r. (The sequence $\{t_r\}$ counts the number of permutations that have only cycles of length 1 or 2. Such permutations are called **involutions**. We shall look at them in detail later in the chapter.)

Exercise 7.8

Prove that $\sum_{k\geq 0} \binom{r}{k} L_k L_{r-k} = 2^r L_r + 2$.

Exercise 7.9

By Theorem 5.7, we know that the terms of the push-button lock sequence $\{c_r\}$ are given by

$$c_r = \sum_{k=0}^{r} \binom{r}{k} k!.$$

Use the product formula to convert this into an *egf*.

Exercise 7.10

The *egf* of the sequence $\{u_r\}$ which counts the number of ways of choosing k elements of the set $\{1, 2, \cdots, r\}$ and arranging the chosen elements into two lists (ordered sets) is given by

$$\sum_{r\geq 0} \frac{u_r z^r}{r!} = \frac{e^z}{(1-z)^2}.$$

Prove that

$$u_r = \sum_{k\geq 0} \binom{r}{k} (k+1)!.$$

7.4 Counting Fixed Points in Permutations

We shall see that the fixed points in permutations are dependent on the derangement numbers. Derangements are the permutations that have no fixed points – that is, no element is mapped to itself. When the permutation is written in cycle form, the fixed points stand out as **singleton** cycles. In Example 1.27 and Equation (1.2) (page 14) we found an explicit form for the derangement numbers:

$$d_r = r! \sum_{k=0}^{r} \binom{r}{k} (-1)^k (r-k)! = r! \sum_{k\geq 0} \frac{(-1)^k}{k!}.$$

You were also invited to derive this from the *egf* in Exercise 7.6. This leads to an approximation and a surprising consequence.

Example 7.18

We compare this explicit form of the derangement numbers with an exponential and derive another form for them. We have

$$d_r = r! \sum_{k=0}^{r} \frac{(-1)^k}{k!} = r! \left(\sum_{k=0}^{\infty} \frac{(-1)^k}{k!} - \sum_{k=r+1}^{\infty} \frac{(-1)^k}{k!} \right)$$

$$= r!e^{-1} - r! \sum_{k=r+1}^{\infty} \frac{(-1)^k}{k!}.$$

The final term on the right depends on r, and we denote it R_r. It takes the value

$$R_r = r! \sum_{k=r+1}^{\infty} \frac{(-1)^k}{k!} = \frac{(-1)^{r+1}}{r+1} + \frac{(-1)^{r+2}}{(r+1)(r+2)} + \cdots$$

$$= (-1)^{r+1} \left(\frac{1}{r+1} - \frac{1}{(r+1)(r+2)} + \frac{1}{(r+1)(r+2)(r+3)} - \cdots \right).$$

More importantly, the size of R_r is such that

$$|R_r| < \left(\frac{1}{r+1} + \frac{1}{(r+1)^2} + \frac{1}{(r+1)^3} + \cdots \right) = \frac{1}{r}.$$

So we now have

$$d_r = \frac{r!}{e} - (-1)^{r+1} R_r = \frac{r!}{e} + (-1)^r R_r$$

and since, $|R_r| < \frac{1}{r}$, we have the approximation

$$d_r \approx \frac{r!}{e}$$

which is accurate to the nearest positive integer.

We now collect together the results for derangement numbers.

Theorem 7.19 (The derangement numbers)

The derangement numbers $\{d_r\}$ are such that:

(i) they obey the recurrences

$$d_r = r(d_{r-1} + d_{r-2}), d_r = rd_{r-1} + (-1)^r \text{ and } r! = \sum_{k \geq 0} \binom{r}{k} d_k;$$

(ii) they have the explicit form

$$d_r = r! \sum_{k=0}^{r} \frac{(-1)^k}{k!}$$

which may also be written

$$d_r = \frac{r!}{e} + (-1)^r R_r$$

in which $|R_r| < \frac{1}{r}$ and $R_r \to 0$ as $r \to \infty$;

(iii) they have the generating function

$$\sum_{r \geq 0} \frac{d_r z^r}{r!} = \frac{e^{-z}}{1-z};$$

(iv) they have the initial values $\{1, 0, 1, 2, 9, 44, 265, \ldots\}$.

Example 7.20

We use the approximate form to find values for d_5 and d_6. We have:

(i) $d_5 = \frac{5!}{e} - R_5 = 44.14553294\ldots - R_5$
in which $|R_5| < \frac{1}{5} = 0.2$. Hence, $d_5 = 44$;

(ii) $d_6 = \frac{6!}{e} + R_6 = 264.8731976\ldots + R_6$
in which $|R_6| < \frac{1}{6} = 0.166$. Hence, $d_6 = 265$.

Example 7.21

We can easily find the probability that a random permutation has no fixed points. There are $r!$ permutations of r objects and there are

$$\frac{r!}{e} + (-1)^r R_r$$

permutations with no fixed points. So the probability (P_r) that a random permutation of r objects is a derangement is

$$P_r = \frac{1}{e} + (-1)^r \frac{R_r}{r!}.$$

As $r \to \infty$ we have the astonishing result that $P_r \to e^{-1}$.

We conclude with similar results for the push-button lock sequence.

Theorem 7.22 (The push-button lock sequence)

The push-button lock combinations $\{c_r\}$ are such that

(i) they obey the recurrences

$$c_r = rc_{r-1} + 1, c_r = (r+1)c_{r-1} - (r-1)c_{r-2} \text{ and } r! = \sum_{k=0}^{r}(-1)^{r-k}\binom{r}{k}c_k;$$

(ii) they have the explicit form

$$c_r = \sum_{k=0}^{r}\binom{r}{k}k!;$$

(iii) they have the *egf*

$$\sum_{r \geq 0}\frac{c_r z^r}{r!} = \frac{e^z}{1-z}.$$

Proof

These have all been given in examples or as an exercise:

(i) Exercises 5.6, 5.7 and 5.8;

(ii) Theorem 5.7;

(iii) Exercise 7.5.

$\square$

We now investigate the distribution of fixed points in permutations. We begin with the notation used to count the number of permutations (of r objects) that have a given number of fixed points.

Definition 7.23 (Number of permutations with precisely k fixed points)

The number $d_{r,k}$ counts the number of permutations of r objects with precisely k fixed points.

Note: a permutation with no fixed points is a derangement and so with this notation we have two (equivalent) ways of writing the same thing: $d_{r,0} \equiv d_r$.

Theorem 7.24

The explicit form for $d_{r,k}$ is given in terms of derangement numbers $d_{r,k} = \binom{r}{k}d_{r-k}$.

Proof

A permutation with precisely k fixed points may be constructed as in Figure 7.1. $\square$

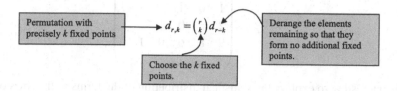

Figure 7.1 Permutations with precisely k fixed points.

Example 7.25

Using this result we may find the distribution of fixed points in the permutations of four objects. There are, by the previous theorem

(i) $d_{4,4} = \binom{4}{4}d_0 = 1.1 = 1$ permutation with four fixed points;

(ii) $d_{4,3} = \binom{4}{3}d_1 = 4.0 = 0$ permutations with three fixed points (if it has three fixed points, then it must have four – this is where the "precisely" comes in);

(iii) $d_{4,2} = \binom{4}{2}d_2 = 6.1 = 6$ permutations with two fixed points;

(iv) $d_{4,1} = \binom{4}{1}d_3 = 4.2 = 8$ permutations with one fixed point;

(v) finally, there are $d_4 \equiv d_{4,0} = \binom{4}{0}d_4 = d_4 = 4! - 1 - 0 - 6 - 8 = 9$ permutations with no ($= 0$) fixed points.

For a given value of r we have a corresponding set of enumerators $d_{r,k}$, one for each value of k. We can arrange these enumerators into an array. We call the array the Fixed point matrix.

Definition 7.26 (The Fixed point matrix)

The Fixed point matrix is the array $S = (d_{r,k})$.

This array graphically displays the distribution of fixed points in permutations:

$$S = (d_{r,k}) = \begin{pmatrix} 1 & & & & & \cdot \\ 0 & 1 & & & & \cdot \\ 1 & 0 & 1 & & & \cdot \\ 2 & 3 & 0 & 1 & & \cdot \\ 9 & 8 & 6 & 0 & 1 & \cdot \\ \cdot & \cdot & \cdot & \cdot & \cdot & \cdot \end{pmatrix}$$

It is interesting to explore the statistical distribution of the terms in the rows of the Fixed point matrix. Can we find the mean of the terms in each row and their variance? These measures give us information about the way that fixed points occur in random permutations. We use group theory to answer these questions.

Example 7.27

Theorem 6.4 gives a neat way to calculate the mean of the number of fixed points that a random permutation has. We apply Equation (6.1) (page 109) to $\mathrm{Sym}(n)$. Note that $\mathrm{Sym}(n)$ has just one orbit, so

$$1 = \frac{1}{n!} \sum_{g \in \mathrm{Sym}(n)} |\mathrm{Fix}(g)|.$$

So the mean is just 1.

Now using Corollary 6.2 we can calculate the variance. We observe that $(\#\mathrm{Fix}_G(g))^2$ is the number of fixed points that the element g has on ordered pairs. So when $G = \mathrm{Sym}(n)$, this counts how many of the pairs (r,s), $1 \leq r, s \leq n$ are fixed by g. By Equation (6.2)

$$2 = \frac{1}{|G|} \sum_{g \in \mathrm{Sym}(n)} |\mathrm{Fix}(g)|^2.$$

So the variance is just $2 - 1 = 1$.

Each of these statistical measures is, rather amazingly, independent of r. The distribution of fixed points in permutations aspires to be Poisson(1). It's as if fixed points are accidents waiting to happen!

7.5 Counting more Permutations

We continue to count permutations of particular types. We start with the number of permutations that contain cycles only of a specified type.

7.5.1 Counting Involutions

Definition 7.28 (Involutions)

A permutation made up only from cycles of length 1 or 2 is called an involution.

Note: such permutations are always self inverse.

Definition 7.29 (Number of permutations of a particular type)

The number of permutations of r objects of a particular type will be written in two ways: in the "long" form $\sigma(r, \text{type})$ and sometimes in the concise form $\sigma_{r,\text{type}}$.

Example 7.30

Using this notation, we must write the involutions in this way:

$$\sigma(r : \text{cycles have length} \leq 2) \text{ or as } \sigma_{r,\leq 2}.$$

We can easily find some initial values:

(i) of the permutations of $\{1\}$ there is 1 whose cycles are of the required type. So

$$\sigma(1 : \text{cycle length} \leq 2) = 1;$$

(ii) of the permutations of $\{1,2\}$ there are 2 whose cycles are of the required type. So

$$\sigma(2 : \text{cycle length} \leq 2) = 2;$$

(iii) the permutations of $\{1,2,3\}$ are $(1)(2)(3), (12)(3), (13)(2), (23)(1), (123)$ and (132). Of these there are four (the first four) whose cycles are of the required type. So

$$\sigma(3 : \text{cycle length} \leq 2) = 4.$$

So the involution sequence begins

$$\{\sigma(r : \text{cycle length} \leq 2)\} = \{1, 1, 2, 4, \cdots\}.$$

We can find a generating function for the sequence of involutions.

Theorem 7.31

The *egf* of the involutions is given by

$$\sum_{r \geq 0} \frac{\sigma(r, \text{cycle length} \leq 2) z^r}{r!} = e^{z + \frac{z^2}{2}}.$$

Proof

An involution has *either* 0 fixed points and all its other cycles of length 2; *or* 1 fixed point and all its other cycles of length 2, and so on. This means we can construct involutions as in Figure 7.2.

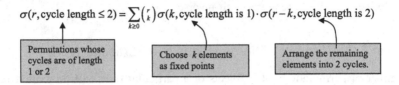

Figure 7.2 Involutions.

If we look carefully at the formula in the figure it is seen to be in the form of Theorem 7.12 – that is, as the product of two *egf*s:

(i) the first part of the product is the *egf* of the number of permutations which are made up only from fixed points. There is always only one such permutation (the identity) and so

$$\sum_{r \geq 0} \frac{\sigma(r, \text{cycle length is } 1) z^r}{r!} = \sum_{r \geq 0} \frac{z^r}{r!} = e^z;$$

(ii) the second part of the product is the *egf* of the number of permutations all of whose cycles are of length 2. But we know how to count these (Theorem 4.18)

$$\sigma(r : \text{cycle length is } 2) = \begin{cases} 0 \text{ if } r \text{ is odd}; \\ \frac{(2m)!}{2^m m!} \text{ when } r = 2m. \end{cases}$$

This means that its *egf* is

$$\sum_{r \geq 0} \frac{\sigma(r : \text{cycle length is } 2) z^r}{r!} = \sum_{s \geq 0} \frac{(2s)!}{2^s s!} \frac{z^{2s}}{(2s)!} = \sum_{s \geq 0} \frac{(z^2)^s}{2^s s!} = e^{\frac{z^2}{2}}.$$

Finally, putting this all together, we have

$$\sum_{r \geq 0} \frac{\sigma(r, \text{cycle length} \leq 2) z^r}{r!} = e^z \cdot e^{\frac{z^2}{2}} = e^{z + \frac{z^2}{2}}.$$

$\square$

From this it is easy to derive a recurrence as in the following example.

Example 7.32

We simply differentiate the *egf* (using concise notation):

$$\sum_{r\geq 0} \frac{\sigma_{r,\leq 2} z^r}{r!} = e^{z+\frac{z^2}{2}}$$

which gives

$$\sum_{r\geq 0} \frac{\sigma_{r+1,\leq 2} z^r}{r!} = (1+z)e^{z+\frac{z^2}{2}} = (1+z)\sum_{r\geq 0}\frac{\sigma_{r,\leq 2} z^r}{r!}$$

and, if we compare coefficients of z^r on either side, then

$$\frac{\sigma_{r+1,\leq 2}}{r!} = \frac{\sigma_{r,\leq 2}}{r!} + \frac{\sigma_{r-1,\leq 2}}{(r-1)!}$$

and hence the required result

$$\sigma_{r+1,\leq 2} = \sigma_{r,\leq 2} + r\sigma_{r-1,\leq 2}.$$

There is an enumerative argument for the same recurrence.

Example 7.33

The number of involutions of $r-1$ objects is $\sigma_{r-1,\leq 2}$. If we introduce a new element r there will be $\sigma_{r,\leq 2}$ involutions. However, in introducing the new element there are two possibilities for it: *either* it makes its own cycle (so it becomes a fixed point), *or* it takes one of the existing fixed elements and together they make a 2 cycle as in Figure 7.3. Hence, overall

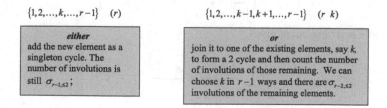

Figure 7.3 Constructing involutions.

$$\sigma_{r,\le2} = \sigma_{r-1,\le2} + (r-1)\sigma_{r-2,\le2}.$$

Using this recurrence with the initial values $\sigma_{0,\le2} = 1$ and $\sigma_{1,\le2} = 1$ the sequence continues

$$\{\sigma_{r,\le2}\} = \{1,1,2,4,10,26,\cdots\}.$$

This leads to a very natural companion to this result – counting the number of permutations, each of whose cycles are of length 3 or more.

Example 7.34

The number of such permutations is $\sigma(r : \text{cycle length} > 2) = \sigma_{r,>2}$. In Figure 7.4 we construct the totality of permutations by bringing together the two different types of permutation (involutions with those whose cycles are of length 3 or more).

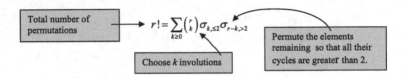

Figure 7.4   Constructing permutations.

We may

(i) choose k elements from r objects in $\binom{r}{k}$ ways and there are $\sigma_{k,\le2}$ permutations of these elements, whose cycles are all of length 2 or less;

(ii) there are $\sigma_{r-k,>2}$ permutations of the remaining elements all of whose cycles are of length greater than 2.

Now we appeal to the product formula (Theorem 7.12) and then we have:

$$\frac{1}{1-z} = \left(\sum_{r\ge0}\frac{\sigma_{r,\le2}z^r}{r!}\right)\left(\sum_{r\ge0}\frac{\sigma_{r,>2}z^r}{r!}\right)$$

and a simple rearrangement gives

$$\sum_{r\ge0}\frac{\sigma_{r,>2}z^r}{r!} = \frac{e^{-z-\frac{z^2}{2}}}{1-z}.$$

The number of permutations all of whose cycles are of length 3 or more satisfies a recurrence. You will be invited to derive this from the *egf* in Exercise 7.19. We give an enumerative derivation – with a vital dash of algebra.

Example 7.35

We take as the elements to be permuted $\{1, 2, \ldots, r\}$. There are $\sigma_{r,>2}$ such permutations. The element 1 belongs to a cycle of length 3 or more – so it belongs to a cycle of length 3, *or* 4, *or* 5, and so on:

(i) suppose it belongs to a 3 cycle. There are $(r-1)(r-2)\sigma_{r-3,>2}$ such permutations, since there are $(r-1)(r-2)$ 3 cycles containing the element 1, while the remaining elements can be permuted in $\sigma_{r-3,>2}$ ways;

(ii) next, suppose the element 1 belongs to a 4 cycle. There are $(r-1)(r-2)(r-3)\sigma_{r-4,>2}$ such permutations.

If we continue this argument, then we have:

$$\sigma_{r,>2} = (r-1)(r-2)\left[\sigma_{r-3,>2} + (r-3)\sigma_{r-4,>2} + (r-3)(r-4)\sigma_{r-5,>2} + \cdots\right]$$
$$= (r-1)(r-2)\sigma_{r-3,>2} + (r-1)\left[(r-2)(r-3)\sigma_{r-4,>2} + \cdots\right]$$
$$= (r-1)(r-2)\sigma_{r-3,>2} + (r-1)\sigma_{r-1,>2}$$

and so the required recurrence is

$$\sigma_{r,>2} = (r-1)\sigma_{r-1,>2} + (r-1)(r-2)\sigma_{r-3,>2}.$$

Using the recurrence with the initial values $\sigma_{0,>2} = 1$, $\sigma_{1,>2} = 0$, and $\sigma_{2,>2} = 0$ the sequence continues

$$\{\sigma_{r,>2}\} = \{1, 0, 0, 2, 6, 24, \ldots\}.$$

7.5.2 Counting Zig-Zag Permutations

We can also count permutations whose image – in the bijective form – has a particular form, just as we counted ascents in Chapter 5. This next example concerns permutations whose image alternately rises and then falls. It has a very surprising *egf*.

Definition 7.36 (Zig-zag permutations)

The zig-zag permutations are those that begin with a rise and thereafter alternately fall followed by a rise and so on.

Example 7.37

The permutation $(1)(23)(45)$ may be written in bijective form

$$\begin{pmatrix} 12345 \\ 13254 \end{pmatrix};$$

this is a zig-zag permutation, since the image goes $1 \nearrow 3 \searrow 2 \nearrow 5 \searrow 4$ as required.

Example 7.38

Now we construct a recurrence satisfied by the zig-zag numbers. In counting such permutations we observe that the **odd** and **even** cases are different, as Figure 7.5 shows.

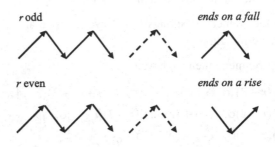

Figure 7.5 Zig-zag permutations.

We deal with the **odd** case first. Suppose that there are o_r zig-zag permutations of the set $\{1,2,3,\ldots,r\}$ and r is even. We may introduce an additional element $r+1$ which is then odd. We can construct the zig-zag permutations of the enlarged set $\{1,2,3,\ldots,r,r+1\}$, and there are o_{r+1} of them, by placing this (largest) element amongst the elements of the original set. It may be placed at any position but the first and last – these would give a descent and an ascent respectively. Further, it must be placed at an even indexed position. Then we construct the zig-zag permutations as in Figure 7.6. So now we have

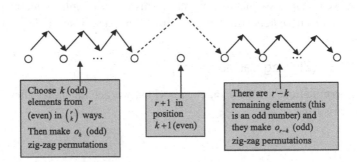

Figure 7.6 Odd zig-zag permutation.

$$o_{r+1} = \sum_{k \text{ odd}} \binom{r}{k} o_k o_{r-k}.$$

However, if we agree that $o_r = 0$ whenever r is even (including zero), then we can write this (for $r \geq 1$)

$$o_{r+1} = \sum_{k>0} \binom{r}{k} o_k o_{r-k}, \text{ with } o_1 = 1.$$

If we consider the even case e_r and agree that $e_r = 0$, whenever r is odd, we find by a similar enumeration that

$$e_{r+1} = \sum_{k>0} \binom{r}{k} o_k e_{r-k}.$$

The latter recurrence is "mixed". However, both recurrences may be used to derive their respective *egf*s.

Theorem 7.39 (*Egf* of the zig-zag numbers)

The *egf* of the zig zag numbers $\{a_r\}$ is

$$Z(z) = O(z) + E(z) = \sum_{r \geq 0} \frac{a_r z^r}{r!} = \tan z + \sec z.$$

Proof

We start with the first recurrence. Again we recognize the result of a product of *egf*s and making use of Theorem 7.12 gives

$$\sum_{r \geq 0} \frac{o_{r+1} z^r}{r!} = \left(\sum_{r \geq 0} \frac{o_r z^r}{r!} \right) \left(\sum_{r \geq 0} \frac{o_r z^r}{r!} \right) + 1$$
$$= (O(z))^2 + 1.$$

But the term on the left (from Theorem 7.9) may be rewritten as a derivative and then we have a differential equation with a unique solution

$$O'(z) = (O(z))^2 + 1$$
$$\Rightarrow O(z) = \tan z.$$

Working in the same way on the second recurrence gives $E(z) = \sec z$. □

7.5.3 Exercises

Exercise 7.11

Show that the sequence $\{\sigma_{r,=2}\}$, which counts the number of permutations of r elements that consist only of cycles of length 2:

(i) satisfies the recurrence $\sigma_{r+1,=2} = r\sigma_{r-1,2}$

(ii) deduce that $\sigma_{2r,=2} = 1.1.3.\cdots.(2r-1)$.

Exercise 7.12

Show that:

(i) the *egf* of the sequence $\{\sigma(r : \text{all cycles of length 3})\}$ which counts the number of permutations whose cycles are all of length 3 is

$$\sum_{r \geq 0} \frac{\sigma_{r,=3} z^r}{r!} = e^{\frac{z^3}{3}} ;$$

(ii)

$$\sigma_{r,\leqslant 3} = \sum_{k \geqslant 0} \binom{r}{k} \sigma_{k,=3} \sigma_{r-k,\leqslant 2}.$$

Deduce that the *egf* of the sequence $\{\sigma(r : \text{cycles of length} \leq 3)\}$ is

$$\sum_{r \geq 0} \frac{\sigma_{r,\leq 3} z^r}{r!} = e^{z + \frac{z^2}{2} + \frac{z^3}{3}} .$$

We could continue this process – of which sequence is the function

$$e^{z + \frac{z^2}{2} + \frac{z^3}{3} + \frac{z^4}{4} + \cdots}$$

the *egf*?

Exercise 7.13

By differentiating the *egf* of the sequence $\{\sigma(r : \text{all cycles of length 2})\}$ twice, show that the terms of this sequence also satisfy the recurrence $\sigma_{r+2,=2} = \sigma_{r,=2} + r(r-1)\sigma_{r-2,=2}$.

7.6 Miscellaneous Exercises

Exercise 7.14

Find the ratio (as $r \to \infty$) of the number of push-button lock combinations $\{c_r\}$ to the number of permutations.

Exercise 7.15

Show that the *egf* of the sequence $\{\sigma(r : \text{no cycles of length 2})\} \equiv \{\sigma_{r,\neq 2}\}$ is

$$\sum_{r \geq 0} \frac{\sigma_{r,\neq 2} z^r}{r!} = \frac{e^{-\frac{z^2}{2}}}{1-z}.$$

By differentiating this, show that the sequence satisfies the recurrence

$$\sigma_{r+1,\neq 2} = (r+1)\sigma_{r,\neq 2} - r\sigma_{r-1,\neq 2} + r(r-1)\sigma_{r-2,\neq 2}.$$

Exercise 7.16

Establish the identity

$$c_r = \sum_{k \geq 0} \binom{r}{k} 2^{r-k} d_k,$$

in which $\{c_r\}$ is the push-button lock sequence and $\{d_r\}$ is the derangement sequence.

Exercise 7.17

The two recurrences for the (odd/even) zig-zag numbers can be used to find the initial numbers of these permutations. Given that

(i) $o_0 = 0$ and $o_1 = 1$, show that the odd zig-zag numbers continue $\{o_r\} = \{0, 1, 0, 2, 0, 16, 0, 272, \ldots\}$;

(ii) $e_0 = 1$ and $e_1 = 0$, show that the even zig-zag numbers continue $\{e_r\} = \{1, 0, 1, 0, 5, 0, 61, \ldots\}$.

Exercise 7.18

The sequence $\{u_r\}$ has (a **sort** of) *egf*

$$\sum_{r \geq 0} \frac{u_{r+1} z^r}{r!} = \frac{e^{-z}}{(1-z)^2}.$$

Prove that it satisfies the two recurrences:

(i) $u_{r+1} = ru_r + d_r$ where $\{d_r\}$ is the derangement sequence;

(ii) $u_{r+2} = (r+1)u_{r+1} + ru_r$.

Find the first five terms of the sequence. (This sequence counts the number of permutations of $\{1, 2, \ldots, r\}$ that have no sub-strings of the form $\ldots k, k+1, \ldots$ for any k, in which $0 \leq k \leq r-1$.)

Exercise 7.19

Show (from the *egf*) that the recurrence satisfied by the number of permutations that are not involutions $\sigma_{r,>2}$ is

$$\sigma_{r,>2} = (r-1)\sigma_{r-1,>2} + (r-1)(r-2)\sigma_{r-3,>2}.$$

Exercise 7.20

Use the recurrence of Example 7.38 for the even zig-zag numbers

$$e_{r+1} = \sum_{k=0}^{r} \binom{r}{k} o_k e_{r-k}$$

to derive its *egf*.

8

Graphs

8.1 Introduction to Graphs

In this chapter we will introduce the idea of graphs, but these are not the pictures with axes and lines which you may have been used to. This use of the word graph represents a collection of vertices and of edges. They are usually shown as very simple pictures, where the vertices are represented by dots and edges by lines.

We begin by drawing some simple graphs – see Figure 8.1. Then we will give a more formal definition of what we mean. In these examples one can see that the vertices are $\{a,b,c\}$. Each graph has the same set of vertices but different sets of edges. So to define a graph we need to say what the vertices are and what the edges are.

Definition 8.1 (Graph)

A graph Γ is defined as a pair (V, E) where E is a subset of unordered pairs of elements from V. V is called the set of vertices and E is called the set of edges. If $e = (u, v)$ is an edge then the vertices u and v are the endpoints of e.

A graph satisfying this definition is sometimes called a simple graph, as we allow no loops, that is edges which begin and end at the same vertex. We also do not allow multiple edges: that is, between two vertices u and v there is at most one edge. One way to define a graph is just to list the edges. Table 8.1 gives the lists for the graphs shown in Figure 8.1:

A. Camina, B. Lewis, *An Introduction to Enumeration*,
Springer Undergraduate Mathematics Series,
DOI 10.1007/978-0-85729-600-9_8, © Springer-Verlag London Limited 2011

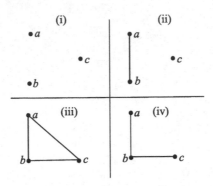

Figure 8.1 Some labelled graphs on three vertices.

Graph (i)	Graph (ii)	Graph (iii)	Graph (iv)
	(a,b)	(a,b) (b,c)	(a,b) (b,c)
			(a,c)

Table 8.1 List for the graphs in Figure 8.1.

Note that the first column is empty because there are no edges! Clearly this is not a sensible way to describe graphs of large size, but it has some uses in computer representations of graphs. Drawing pictures is also very useful, but again one has to be careful since one can draw a particular graph in many different ways. So we need an idea as to when two graphs are the same. But before that we need to discuss the difference between a labelled graph and an unlabelled graph. The way that the graphs have been drawn in Figure 8.1 is as labelled graphs. In two of the graphs, that with no edges (i) and that with three edges (iii), it clearly makes no difference how we label the graphs. There are, however, three different ways to label the other two graphs, so we see that we can get eight graphs which are labelled and seem to be different (see Figures 8.2 and 8.1).

Definition 8.2 (Isomorphism of graphs)

Let $\Gamma_1 = (V_1, E_1)$ and $\Gamma_2 = (V_2, E_2)$ be two graphs. We say that Γ_1 and Γ_2 are isomorphic if there is a bijection between V_1 and V_2 which preserves edges.

Definition 8.3 (Degree of vertex)

Let $\Gamma = (V, E)$ be a graph then for each $v \in V$ the degree of v is the number of edges

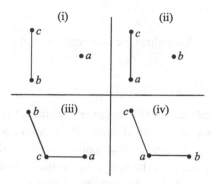

Figure 8.2 Four labelled graphs on three vertices.

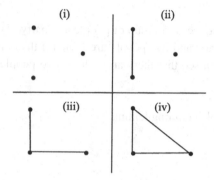

Figure 8.3 (Un)labelled graphs on three vertices.

whose endpoints include v. We will denote the degree of a vertex by $d(v)$.

There are two graphs which can be defined on any set V: the **null graph**, which has no edges, and the complete graph, in which any two vertices are joined by an edge. If $|V| = r$ then the null graph has 0 edges and the degree of each vertex is 0. The **complete graph** has $\binom{r}{2}$ edges and each vertex has degree $r - 1$. Let $\Gamma_1 = (V_1, E_1)$ and $\Gamma_2 = (V_2, E_2)$ be two graphs and let ϕ be an isomorphism between them. Then for each $v \in V$ we note that both v and $\phi(v)$ must have the same degrees. This is one of the easiest ways to show that two graphs are **not** isomorphic. Hence it is easy to see that the four unlabelled graphs in Figure 8.3 are all non-isomorphic. However, this offers very little help when trying to decide that two graphs are isomorphic.

Lemma 8.4

Let $\Gamma = (V, E)$ and $|V| = r$. Then there are two vertices with the same degree.

Proof

There are r vertices and the degrees have to come from the set $\{0, 1, \ldots, r-1\}$. If there is a vertex of degree 0 there is not one of degree $r-1$. So we are choosing r numbers from a set with only $r-1$ elements. So we know two of the degrees have to be the same by the pigeonhole principle (Theorem 1.12). If there are no vertices the degrees come from the set $\{1, \ldots, r-1\}$ and the same proof works. □

Example 8.5

An intriguing example concerns a group of people at a party. We draw the graph Γ so that each person is a vertex and two people are joined if they know each other. From the above lemma one can see that there are at least two people who know the same number of people.

Next we have a simple counting lemma.

Lemma 8.6

Let $\Gamma = (V, E)$ be a graph and let $|E| = m$. Then

$$\sum_{v \in V} d(v) = 2m.$$

Proof

Just note that each edge has exactly two vertices and each vertex v will be counted exactly $d(v)$ times. □

A simple consequence of this lemma is that there must be an even number of vertices with odd degree otherwise the sum of the degrees has to be odd.

Example 8.7

Another example is to consider road systems in a country. We may construct a graph defining the towns as the vertices and the roads as the edges connecting them. In this case we have to assume that there is at most one road connecting any two towns so as to make the graph simple. To decide whether one can travel from one town to another we only need to know that there exists at least one route. We might also consider railway lines, flight connections, rivers or canals rather than roads.

This example motivates the following definition.

Definition 8.8 (Paths in a graph)

Let $\Gamma = (V, E)$ be a graph. A path in a graph is a sequence $(v_0, e_1, v_1, \ldots, e_n, v_n)$ where each $v_i \in V$ and each $e_i \in E$ where the endpoints of e_i are v_{i-1} and v_i for $i \in \{0, 1, \ldots, n\}$. The **length** of such a path is n. Given such a path $(v_0, e_1, v_1, \ldots, e_n, v_n)$ we define the following:

(i) the vertices v_0 and v_n are said to be **connected**;

(ii) a path is called a cycle if $v_0 = v_n$;

(iii) a simple path is a path in which $e_i \neq e_j \forall 1 \leq i, j \leq n$;

(iv) a simple closed path or simple closed cycle is a simple path which is a cycle.

If loops and multiple edges are allowed we have to be more careful with the definitions we adopt.

Lemma 8.9

Let $\Gamma = (V, E)$ be a graph. The relation "is connected" is an equivalence relation.

Proof

The reflexivity and symmetry are trivial. Let u and v be connected by the path $(v_0, e_1, v_1, \ldots, e_n, v_n)$ so that $u = v_0$ and $v = v_n$ and let v and w the connected by the path $(u_0, f_1, u_1, \ldots, f_m, u_m)$ where $v = u_0$ and $w = u_m$. Now it is easy to see that we can just join the two paths and then u and w are connected. □

Definition 8.10 (Connected component)

The equivalent classes are known as the connected components of the graph and if there is only one connected component the graph is said to be connected.

Clearly the null graph on n vertices has n connected components and the complete graph has just one.

Definition 8.11 (Subgraph)

Let $\Gamma = (V, E)$ be a graph. A subgraph of Γ is a subset of the vertices and edges so that if an edge e is in the subgraph then so are both of its endpoints.

In using induction, it is often useful just to delete an edge or a vertex. If a vertex is deleted, all the edges connected to the vertex also have to be deleted. We will use the notation $\Gamma \setminus x$ to denote such a subgraph, where x is a vertex or an edge.

Definition 8.12 (Bipartite)

A graph $\Gamma = (V, E)$ is said to be bipartite if there exists a partition of V into two disjoint subsets, say U and W such that both U and W are null graphs. The union of the two subgraphs $U \cup W$ is called the bipartition.

Another way to say this is to say that for any edge e one endpoint is in U and the other is in W.

Example 8.13

Given two natural numbers m and n, the **complete bipartite graph**, denoted by $K_{m,n}$, is the graph with $V = U \cup W$ and $E = \{(u, w) : u \in U, w \in W\}$ where $|U| = m$ and $|V| = n$.

Example 8.14

Let U be any set and let W be a collection of subsets of U. The graph $\Gamma = (V, E)$ is defined by $V = U \cup W$ and $E = \{(u, w) : u \in U, w \in W \text{ with } u \in w\}$.

Lemma 8.15

A graph $\Gamma = (V, E)$ is bipartite if and only if it has no closed simple path of odd length.

Proof

Assume $\Gamma = (V, E)$ is bipartite with $V = V_1 \cup V_2$ and let $(v_0, e_1, v_2, \ldots, e_n, v_n)$ be a cycle. Assume without loss of generality that $v_0 \in V_1$. Then $v_1 \in V_2$, $v_2 \in V_1$ and $v_i \in V_{i+1}$. If n is odd then $v_n \in V_2$ and so cannot be equal to v_0.

Now assume that $\Gamma = (V, E)$ has no odd closed simple paths. We prove that if Γ is connected it is bipartite and then it follows if the graph is disconnected by looking at each connected component. For any pair of vertices, v and w say, define $d(v, w)$ to be the length of the shortest path from v to v, let $d(v,v) = 0, \forall v \in V$. Define a relation on V by $v \sim w$ if $d(v,w)$ is even. We need to show that this is an equivalence relation. It is clearly reflexive and symmetric. So we need to consider transitivity. Let $w_1 \sim w_2$ and $w_2 \sim w_3$, that is both $d(w_1, w_2)$ and $d(w_2, w_3)$ are even. Assume that $d(w_1, w_3)$ is odd. If we join the three paths there is a cycle of odd length, but it might have both repeated edges and repeated vertices. Assume that there is a repeated edge and choose the first repeat, $(v_{i-1}, e_i, v_i, \ldots, e_n, v_n)$, where $e_i = e_n$. So $v_{n-1} = v_{i-1}$ and $v_n = v_i$. If $(v_{i-1}, e_i, v_i, \ldots, e_{n-1}, v_{n-1})$ has odd length then this is an odd simple closed path, which is not allowed. Hence the length is even and we can just delete the section and still have a cycle of odd length. So we can assume that there are no repeated edges.

We can now repeat the argument for repeated vertices. If the first repeat is $(v_{i-1}, e_i, v_i, \ldots, e_n, v_n)$ with $v_{i-1} = v_n$ and the length is odd then there is an odd simple closed path. If not we can delete this part an get a shorter odd cycle.

So $\sim$ is an equivalence relation. Let the two equivalence classes be V_1 and V_2. Note that if two vertices are in the same equivalence class there cannot be an edge between them because that would give a path of length 1. □

One last definition in this section.

Definition 8.16 (Complement)

Given a graph, $\Gamma = (V, E)$, the complement to Γ is the graph with the same vertices where two vertices are joined in the complement if and only if they are not joined in Γ.

This is often a nice way to construct new graphs from old.

8.1.1 Exercises

Exercise 8.1

Show that if Γ is a graph such that every vertex has odd degree then Γ has an even number of vertices.

Exercise 8.2

Let $\Gamma = (V, E)$ be a graph and assume that the degree of each vertex is at least 1. Show that if the number of edges is less than the number of vertices there is a vertex of degree 1.

Exercise 8.3

Let $\Gamma = (V, E)$ be a connected graph with at least two vertices. Show that the degree of each vertex is at least 1.

Exercise 8.4

Prove that if there is a path from u to v then there is a simple path from u to v.

Exercise 8.5

Let $\Gamma = (V, E)$ be a bipartite graph with bipartition $V = U \cup W$. Assume that $|V| = n, |E| = m, |U| = r$ and $|W| = s$. Show that

(i) $m \leq rs$,

(ii) $m \leq \frac{n^2}{4}$ and

(iii) describe the graphs for which equality holds.

8.2 Connectivity

We wish to examine the concept of a connected graph, Definition 8.10, more carefully. Can we construct a way of determining how connected a graph is? We do that by turning a connected graph into a disconnected graph – we measure how connected a graph is by how many edges or vertices it is necessary to remove before the graph becomes disconnected. This idea is clearly important in many areas. If you live somewhere where there is only one road out of your village and it gets cut off, you are trapped. When you log on to the web, you hope that if one of the nodes that you need to connect you is offline, there are others which can be used. So we have the following definitions.

Definition 8.17 (Separating Set)

Let $\Gamma = (V, E)$ be a graph then

(i) A subset of vertices or edges is called a separating set if the graph with that subset removed is disconnected.

(ii) A graph is called k-connected if no subset of vertices of size $k - 1$ is a separating set but a subset of size k is.

(iii) A graph is called k-edge connected if no subset of edges of size $k - 1$ is a separating set but a subset of size k is.

It turns out that there are some interesting special cases.

Definition 8.18 (Trees and forests)

(i) A tree is a connected graph with no simple closed paths. Of the graphs shown in Figure 8.2 two of them are trees.

(ii) A forest is a graph with no simple closed paths.

Note: each connected component of a forest is a tree.

Theorem 8.19

Let Γ be a graph with r vertices and k connected components. Then Γ is a forest if and only if Γ has $r - k$ edges.

Proof

We will prove the result by induction on $r + k$, the first cases are easy to compute.

Assume first that Γ has no cycles. We show that there is a vertex of degree 1. Assume the contrary so that every vertex has degree at least 2. Choose any vertex v_0. This is connected to a vertex $v_1 \neq v_0$. Then, if possible, find a vertex v_2 with $v_2 \notin \{v_0, v_1\}$ but (v_1, v_2) is an edge. Repeat this for as long as possible and get a path, $v_0, e_1, v_1, \ldots, e_r, v_r$ where each e_i is an edge and each v_i is vertex which is distinct. However, v_r has degree 2 so it is connected to a vertex, say v by an edge e. This has to be a vertex v_j or we could extend the path. But then we have a simple closed path $v_j, e_j, v_{j+1}, \ldots, v_r, e, v_j$.

Now choose any vertex of degree 1 and note that deleting this vertex we get a graph with $r - 1$ edges and k components and induction completes the argument that the number of edges is $r - k$.

Now assume that Γ has k components and $r - k$ edges. If there is a vertex of degree 0 then, by deleting that vertex, induction gives the correct answer. If not, by Exercise 8.2 there is a vertex of degree 1. Again deleting this vertex gives an inductive argument. □

Lemma 8.20

Let Γ be a connected graph. Then Γ is a tree if and only if every edge is separating.

Proof

Since a tree has exactly $|V| - 1$ edges, deleting any edge leaves a disconnected graph. To do the converse we use induction on the number of edges. It is obviously true if Γ has one edge. So assume that the result is true for any graph with fewer than n edges and assume Γ has r edges. Deleting an edge e leaves two disconnected components, but both will satisfy the hypothesis of the lemma. So both components are trees and then adding e will give a tree. $\square$

Definition 8.21 (Independent paths)

A set of paths in a graph are called independent when they have no vertices in common, except possibly the first and last vertices.

Theorem 8.22 (Menger's Theorem)

Let $\Gamma = (V, E)$ be a graph and let v and w be two connected vertices of V. Let k be the size of the smallest set U of vertices with the property that v and w are not connected in $\Gamma \setminus U$. Let l be the largest number of independent paths from v to w. Then $k = l$.

Proof

Since removing the set U leaves v and w disconnected it is clear that $l \leq k$. We will use induction on $k + |V|$ to show $k \leq l$. So we have to show that there are at least k independent simple paths from v to w. If $k = 1$ it is clear since there is a simple path from v to w. So assume the result is true for all $r < k$.

Let U be a set of vertices with the property that v and w are not connected in $\Gamma \setminus U$ and let $|U| = k$.

Assume that there is a vertex $u \in U$ so that both vu and uw are edges. Now use induction on the graph $\Gamma \setminus \{u\}$. This says that there are $k - 1$ independent simple paths from v to w in $\Gamma \setminus \{u\}$ and we can add the path $v, (vu), u, (uw), w$ to get k independent simple paths.

Now assume that neither v nor w is joined to every vertex in U. For $x \in \{v, w\}$ let Γ_x be the connected component of $\Gamma \setminus U$ containing x. Now construct a new graph Δ_v which is constructed by taking Γ_v and joining U and all appropriate edges and one new vertex w' joining w' to all vertices in U. Note by the assumption w is not connected to every vertex in U and so the graph Δ_v has fewer vertices than Γ. Also U disconnects v and w' so there are k independent simple paths from v to w' and each one has to go through a different vertex in U. Now repeat the whole process, swapping the roles of v and w and construct k independent simple paths from v' to w each one going through a different vertex of U. Now take the path from v to a vertex in U from Δ_v and join it with the path from that vertex to w in Δ_w to get k independent simple paths.

We are now left with the case when either v or w is joined to every vertex in U and we can assume that this is true for any set of size k disconnecting v and w. So assume that v is joined to each vertex in U. Choose one such vertex, say u, and consider the graph Γ' we get by deleting the edge vu. By induction there will be $k - 1$ disjoint simple paths from v to w in Γ', say $\{p_1, p_2, \ldots, p_{k-1}\}$. There is a simple path in Γ from u to v and if this were independent of the $k-1$ paths in Γ' we would have k independent paths. So any path from u to w has to have a vertex in common with some existing independent path. For each path, say $\{q_1, q_2, \ldots, q_r\}$, from u to w assume that it has points $\{v_i : 1 \leq i \leq m\}$ in common with p_i. Now for any $p_j, k-1 \geq j > 1$ choose a vertex on p_i from the set $\{v_i : 1 \leq i \leq m\}$ not equal to either v or w. It is clear that this set of vertices disconnects v and w but has only $k-1$ elements, which is a contradiction. □

Based on this we can prove a theorem known as **Hall's Marriage Theorem** for a reason that will become apparent in an exercise. However, in the original formulation we have the following: X is a set and $\mathcal{A}$ is a family of subsets of X. Is it possible to find a subset Y of X of size $|\mathcal{A}|$ such that for each $A \in \mathcal{A}$ there is a $y \in A \cap Y$. Another way to describe this is to talk about matchings. A **matching** in a graph Γ is a set of disjoint edges. This is the same as independent edges in the sense that any edge can be thought of as a simple path. In particular, in a bipartite graph any matching is a set of independent simple paths connecting the two parts of the graph. If the bipartite graph has parts of size m and n then the maximal size of a matching is $\min(m,n)$. The next result gives a condition for there to be a maximal matching.

It is clear that if we have such a set then for any subset S of $\mathcal{A}$, $|\bigcup_{A \in \mathcal{A}} A| \geq |C|$. The surprising thing is that this is also a sufficient condition. We can turn this into a graph theory problem. We construct a bipartite graph as follows. Set $V_1 = \mathcal{A}$ and $V_2 = X$ and the only edges are given by pairs (A, x), where $x \in A$ and $x \in X$ and $A \in \mathcal{A}$. For any subset of vertices of a graph, say W, we denote by $N(W)$ the set of vertices adjoined to W. Now we have the following theorem from which the result that the condition is sufficient follows.

Theorem 8.23 (Hall's Marriage Theorem)

Let $\Gamma(V, E)$ be a bipartite graph with partition $V_1 \cup V_2 = V$. Then there is a set of disjoint edges of size $|V_1|$ if and only if $|N(U)| \geq |U|$ for all subsets U of V_1.

Proof

The necessity is clear from the argument above. Now join two new vertices to Γ, say v_1 and v_2, so that v_i is joined to all the vertices in V_i and no others, for $1 \leq i \leq 2$. The condition that there are a set of disjoint edges is the same as saying that there are independent paths from v_1 to v_2. By Theorem 8.22 there are not enough disjoint edges for a matching of size $|V_1|$ if there is a separating set of size less than $|V_1|$. So there exists $U_1 \subset V_1$ and $U_2 \subset V_2$ with $|U_1| + |U_2| < |V_1|$ and $V_1 \setminus U_1$ is only joined to vertices in U_2. We are almost there:

$$|N(V_1 \setminus U_1| \leq |U_2| < |V_1 \setminus U_1|.$$

This violates the condition that $|N(U)| \geq |U|$ for all U. So there are $|V_1|$ independent paths form v_1 to v_2 which completes the proof. □

Example 8.24

Another version of this problem is the assignment problem. That is given a number of vacancies in a company and a number of people applying for the jobs, is it possible to assign people to jobs so that every job is filled by someone who has applied for the job? It depends on how many people apply for how many jobs as in Hall's Marriage Theorem.

8.2.1 Exercises

Exercise 8.6

Let Γ be a connected graph. Show that if one edge is deleted then Γ has at most two connected components.

Exercise 8.7

Let Γ be a graph with r vertices. Show that if each vertex has degree greater than $\frac{r-1}{2}$ then Γ is connected.

Exercise 8.8

Let Γ be a graph with r vertices. Show that if the sum of the degrees of two adjacent vertices is greater than $r - 1$ then Γ is connected.

Exercise 8.9

Given a graph $\Gamma(V, E)$ we can construct the line graph $L\Gamma$ which has as vertices E and two edges are joined if and only if they have a vertex in common. Use the line graph to prove the edge version of Theorem 8.22.

Exercise 8.10

Prove that any tree is a bipartite graph.

8.3 Counting Graphs and Trees

We will divide this section into two subsections each devoted to graphs distinguished by labels.

8.3.1 Counting Labelled Graphs and Trees

In this section we discuss some problems concerned with enumerating labelled graphs. We will find there is a considerable difference between labelled graphs and unlabelled graphs. We begin with some easy results.

Proposition 8.25

(i) There are $2^{\binom{r}{2}}$ labelled graphs with r vertices;

(ii) There are $\left(\binom{r}{2} \atop m\right)$ labelled graphs with r vertices and m edges.

Proof

(i) We can choose any subset of the pairs of distinct vertices and join them.

(ii) This time we choose any subset of size m from the pairs of distinct vertices.

□

Let γ_r denote the number of labelled graphs with r vertices. Let $\gamma_{r,m}$ denote the number of labelled graphs with r vertices and m edges. The generating function (see Section 2.1) for $\gamma_{r,m}$, fixed r, is

$$\sum_{m=0}^{\binom{r}{2}} \gamma_{r,m} z^m = (1+z)^{\binom{r}{2}}.$$

Let c_r denote the number of connected labelled graphs with r edges. Let

$$G(z) = \sum_{r=0}^{\infty} \frac{\gamma_n z^r}{r!}, \quad C(z) = \sum_{r=1}^{\infty} \frac{c_r z^r}{r!}$$

be the exponential generating functions (see Chapter 7) for c_n and γ_n respectively.

We want to show how γ_n and c_n are related. To do this we introduce the idea of a **rooted graph**. A graph is rooted if there is a chosen vertex which which will be called the **root**.

Lemma 8.26

The number, c_r, of labelled connected graphs with r vertices is given by

$$c_r = 2^{\binom{r}{2}} - \frac{1}{r} \sum_{k=1}^{r-1} k \binom{r}{k} 2^{\binom{r-k}{2}} c_k. \tag{8.1}$$

Proof

We count rooted graphs, noting that any labelled graph can be rooted at any of its vertices, so the number of rooted connected graphs with r vertices is $r\gamma_r$. Each root vertex lies in some component with say, k, vertices where $1 \le k \le r$. The k vertices can be labelled in any of $\binom{r}{k}$ ways. There are c_k possible graphs for this component and each can have any of its k vertices chosen as the root. The remaining $r - k$ vertices can come from γ_{r-k} possible graphs. Multiplying everything together and then summing over k gives

$$r\gamma_r = \sum_{k=1}^{r} k \binom{r}{k} \gamma_{r-k} c_k. \tag{8.2}$$

But from Proposition 8.25 we know the numbers $\gamma_k = 2^{\binom{k}{2}}$. Substituting this into the equation above gives

$$r2^{\binom{r}{2}} = \sum_{k=1}^{r-1} k \binom{r}{k} 2^{\binom{r-k}{2}} c_k + r c_r.$$

Rearranging this equation gives the lemma. □

This leads immediately to a surprisingly elegant relationship between the two exponential generating functions.

Theorem 8.27

$$G(z) = \exp(C(z)).$$

Proof

Differentiate $\exp(C(z))$ to give $C'(z) \exp(C(z))$. Now compare the coefficients of z^{r-1} in $C'(z)G(z)$ and $G'(z)$. Then using Equation (8.2) we see that the coefficients are the same. So we have the equality $C'(z)G(z) = G'(z)$. Solving this differential equation gives the result, as long as we define $G(0) = 1$ and $C(0) = 0$. □

In this section we are going to count the number of labelled trees, a result that goes back to Cayley. There are many proofs of this result, but we will choose one that uses recurrence. However, in this problem it is better to count forests, as this helps with the recurrence relation.

We will now count the number of labelled forests with r vertices and k components with a specified subset of size k which are in different components. We will let this number be $t_{r,k}$. By being labelled we mean that each vertex has a label.

Theorem 8.28

$$t_{r,k} = kr^{r-k-1}$$

Proof

Let A be the specified subset and let $v \in A$ have degree i. Deleting v and adding the i vertices connected to it to create a new set A' gives a set of size $k - i + 1$. There are $t_{r-1,k-1+i}$ possible labelled forests that can be constructed with these parameters.

Do this for each possible set of size i from $r - k$ vertices not in A and get the following recurrence relation:

$$t_{r,k} = \sum_{0}^{r-k} \binom{r-k}{i} t_{r-1,k-1+i}.$$

The inductive hypothesis is

$$t_{r,k} = \sum_{0}^{r-k} \binom{r-k}{i} (k-1+i)(r-1)^{r-1-k-i}.$$

Putting $j = r - k - i$

$$= \sum_{0}^{r-k} \binom{r-k}{j} (r-1-j)(r-1)^{j-1}$$

$$= \sum_{0}^{r-k} \binom{r-k}{j} (r-1)^{j} + \sum_{1}^{r-k} \binom{r-k}{j} j(r-1)^{j-1}$$

$$= r^{r-k} + (r-k) \sum_{1}^{r-k} \binom{r-k-1}{j-1} (r-1)^{j-1}$$

$$= r^{r-k} + (r-k) \sum_{0}^{r-k} \binom{r-k-1}{j} (r-1)^{j}$$

$$= r^{r-k} + (r-k)r^{r-1-k} = kr^{r-1-k}.$$

Note that to make this work, the following definitions have to be made: $t_{0,0} = t_{r,r} = 0$, and these cause no problems.   □

Corollary 8.29 (Cayley's Theorem)

The number of labelled trees on r vertices is r^{r-2}.

Proof

Note that $t_{r,1}$ is just the number of labelled trees.   □

In the final part of this section we give another proof of Cayley's theorem, this time using the principle of inclusion–exclusion (see Section 1.3). We want to prove the following identity:

$$\sum_{0}^{r-1}(-1)^i\binom{b}{i}(r-i)^m = 0 \tag{8.3}$$

when $m < r$. We are, however, going to prove something slightly stronger.

Proposition 8.30

The number of onto mappings from a set X with m elements to a set Y with r elements is

$$\sum_{0}^{r-1}(-1)^i\binom{b}{i}(r-i)^m = 0$$

Proof

Let F be the set of all functions $\{f : X \to Y\}$. For each element $U \subset Y$ let S_U be the set of functions $f \in F$ such that U is not in the image of f. Let S_0 be the set of functions f in which every $y \in Y$ is in the image of f. Of course S_0 is just the set we wish to count, that is $S_0 = S_\emptyset$. Then by Theorem 1.20 we get

$$\#(S_0) = \#(S) - \sum_{\#(U)=1} \#(S_U) + \sum_{\#(U)=2} \#(S_U) - \cdots + (-1)^r \#(S_Y).$$

But $\#(S_U) = (r-k)^m$ where $\#(U) = k$ since we are just mapping the set X into $Y \setminus U$. Further there are $\binom{r}{k}$ choice for U of size k. So the above equation becomes

$$\#(S_0) = r^m - \binom{r}{1}(r-1)^m + \binom{r}{2}(r-2)^m - \cdots - \binom{r}{r}(r-r)^m.$$

□

Note that when $m < r$ there is no onto mapping from X to Y so we get Equation (8.3).

We now use this to get another proof of Cayley's Theorem.

Theorem 8.31 (Cayley's Theorem)

The number of labelled trees on r vertices is r^{r-2}.

Proof

We now use inclusion–exclusion and induction. Let $\mathcal{T}_r$ be the set of trees on r vertices, which we will label $1, 2, \ldots, r$. We know that every tree has at least one vertex of degree 1 (see proof of Theorem 8.19). If $r > 2$ not all vertices can be of degree 1. For each $1 \leq i_1 < i_2 \cdots < i_k \leq r$ for $1 \leq k \leq r$ let $\mathcal{T}_{i_1, i_2, \ldots, i_k}$ be the set of trees so that each of $i_1, i_2, \ldots, i_k$ is a vertex of degree 1. Then by Theorem 1.20 we get

$$\#(\mathcal{T}_r) = \sum_i \#(\mathcal{T}_i) - \sum_i \#(\mathcal{T}_{i_1 < i_2}) + \cdots + (-1)^r \sum_{i_1 < i_2 \cdots < i_k} \#(\mathcal{T}_{i_1, i_2, \ldots, i_{r-1}}).$$

Now we can choose the k vertices $i_1 < i_2 \cdots 1_k$ in $\binom{r}{k}$ ways, but for each choice we can connect to the remaining $r - k$ vertices in $(r - k)^k$ different ways and those same $r - k$ vertices can be made into a tree in $\#(\mathcal{T}_{r-k})$ ways. So the equation above becomes

$$\#(\mathcal{T}_r) = \binom{r}{1}(r-1)\#(\mathcal{T}_1) - \binom{r}{2}(r-2)\#(\mathcal{T}_2) + \cdots + (-1)^r \binom{r}{r-1}\#(\mathcal{T}_{r-1})$$

$$= \sum_1^{r-1} (-1)^{k-1} \binom{r}{k}(r-k)^k \#(\mathcal{T}_{r-k})$$

We now use induction and get

$$\#(\mathcal{T}_r) = \sum_1^{r-1} (-1)^{k-1} \binom{r}{k}(r-k)^k (r-k)^{r-k-2}$$

$$= \sum_1^{r-1} (-1)^{k-1} \binom{r}{k}(r-k)^{r-2}.$$

This is just Equation (8.3) with $m = r - 2$. $\qquad\square$

8.3.2 Counting Unlabelled Graphs

Example 8.32

A much harder question is how to count labelled graphs. One way to do this is to think of a graph as a colouring of the pairs (u, v) with $u \neq v$ with S_n acting (see Theorem 6.27). So we need to evaluate the cycle index for $\mathrm{Sym}(n)$ acting on distinct pairs. For any given n this is possible but the general example is messy.

Let us consider the case of $\mathrm{Sym}(4)$. Table 8.2 represents the action.

Element	Cycle type	# of this type
(1)(2)(3)(4)	1^6	1
(12)	$1^2 2^2$	6
(123)	3^2	8
(12)(34)	$1^2 2^2$	3
(1234)	$2^1 4^1$	6

Table 8.2 The symmetric group of degree 4 acting on pairs.

So the cycle index is:

$$P(X_1, X_2, X_3, X_4) = \frac{1}{24}(X_1^6 + 6X_1^2 X_2^2 + 8X_3^2 + 3X_1^2 X_2^2 + 6X_2 X_4).$$

If we replace each X_i with 2 we count the number of graphs on four vertices. So we have 11 graphs on four vertices. If, however, we want to see how many different graphs there are with a given number of edges we have to be more careful. So colour the edges white if they are to be part of the graph and black if we want to excluded them. We can then count using the cycle index to get the following expression (see Theorem 6.27):

$$\frac{1}{24}\left((B+W)^6 + 9(B+W)^2(B^2+W^2)^2 + 8(B^3+W^3)^2\right)$$
$$+ \frac{1}{24}\left(6(B^2+W^2)(B^4+W^4)\right). \quad (8.4)$$

If we look for the coefficient of $B^3 W^3$ we find that it is 3, so there are three unlabelled graphs on four vertices with three edges.

If we consider this example further we can see how to obtain the generating function for the number of graphs on four vertices and i edges. Substitute 1 for B and z for W in Equation (8.4) (page 168), and get:

$$1 + z + 2z^2 + 3z^3 + 2z^4 + z^5 + z^6.$$

Theorem 8.33

Let $\Gamma(V, E)$ be a graph and let G be the group of automorphisms on the vertices of Γ. Then we can label Γ in

$$\frac{|V|!}{|A|}$$

ways.

Proof

Note that no non-trivial element of G will fix a labelled graph, since any change of the labelling will give a new labelling. Also any graph can be labelled in $|V|!$ ways. So by Theorem 6.4 it follows that the number of orbits of G on the set of labelled graphs is $\frac{|V|!}{|A|}$. $\qquad\square$

We now look at an analogue to Theorem 8.27 for (un)labelled graphs. Let κ_n be the number of connected graphs with n vertices and let g_n be the number of graphs with n vertices. Let

$$K(x) = \sum_{n=0}^{\infty} \kappa_n z^n \text{ and } G(x) = \sum_{n=0}^{\infty} g_n z^n$$

be the ordinary generating functions. We wish to relate these but first we prove a lemma.

Lemma 8.34

The number of graphs with kr vertices in which each connected component has r vertices is given by

$$\sum \binom{\kappa_r}{k_0, k_1, k_2, \ldots, k_{\kappa_r}},$$ (8.5)

where the sum is over all $k_0 + k_1 + k_2 + \cdots + k_{\kappa_r} = \kappa_r$ and $k_1 + 2k_2 + \cdots + rk_{\kappa_r} = k$.

Proof

We know that there are κ_n possible connected graphs with r vertices and we need to choose k of them. Recall that we can use repeats, but that the numbers of ways we can choose will depend on how many are the same. So if we choose all the same there are just κ_r choices, but if we have $k - 1$ the same we can choose one of them in κ_r ways and the second in $\kappa_r - 1$ ways.

We describe the choices this way: k_0 are not chosen, k_1 are chosen once, and in general k_i are chosen i times $0 \le i \le \kappa_n$, where

$$\sum_{i=0}^{\infty} k_i = \kappa_r \text{ and } \sum_{i=1}^{\infty} ik_i = k.$$

But if we think there are $\kappa_r!$ choices we get $k_0! k_1! \cdots k_{\kappa_r}!$ repeats. So the number of choices is the multinomial coefficient $\binom{\kappa_r}{k_0, k_1, \cdots k_{\kappa_r}}$. This is similar to the result in Theorem 5.1. $\qquad\square$

The multinomial gets its name because it is the coefficient of a power of z when there are more than two terms. The same argument as above shows that in the expansion of $(z_1 + z_2 + \cdots + z_k)^r$ the coefficient of the term $z_1^{t_1} z_2^{t_2} \cdots z_k^{t_k}$ is $\sum \binom{r}{t_1, t_2, \ldots, t_r}$ where the sum is over all k-tuples with $t_1 + t_2 + \cdots + t_k = r$.

Theorem 8.35

Let $G(z)$ be the generating function for graphs and $K(z)$ be the generating function for connected graphs. Then

$$G(z) = \exp\left(\sum_1^\infty \frac{K(z^k)}{k}\right)$$

Proof

By the preceding lemma and the remark above the generating function for graphs so that each connected component has r vertices is given by

$$(1 + x^r + x^{2r} + x^{3r} + \cdots)^{\kappa_r}.$$

Note that a graph with all connected components with r vertices has kr vertices for some k. Now consider the coefficient of x^{kr} in the above expansion. It will be precisely the sum:

$$\sum \binom{\kappa_r}{k_0, k_1, k_2, \ldots, k_{\kappa_r}},$$

where the sum is over all $k_1 + k_2 + \cdots + k_r = \kappa_r$ and $k_1 + 2k_2 + \cdots + rk_{\kappa_r} = k$.
Now

$$(1 + z^r + z^{2r} + z^{3r} + \cdots)^{\kappa_r} = (1 - z^r)^{-\kappa_r}.$$

Now any graph will be the union of its connected components. We collect together those of the same size and then multiply the generating functions for each set. So

$$G(z) = \prod_{r=1}^\infty ((1 - z^r)^{-\kappa_r})$$

$$= \exp\left(\sum_{r=1}^\infty -\kappa_r \log(1 - z^r)\right)$$

$$= \exp\left(\sum_{r=1}^\infty \sum_{k=1}^\infty \frac{\kappa_r z^{rk}}{k}\right).$$

Now change the order of summation. This is valid since in the expression inside the exp the coefficient of any power of z is a finite sum:

$$= \exp\left(\sum_{k=1}^\infty \left(\sum_{r=1}^\infty \frac{\kappa_r z^{rk}}{k}\right)\right).$$

For each k

$$G(z) = \exp \sum_{k=1}^{\infty} \frac{1}{k} \left(\sum_{r=1}^{\infty} \kappa_r(z^k)^r \right) = \exp \sum_{1}^{\infty} \left(\frac{K(z^k)}{k} \right).$$

□

To end this section we look at an example of unlabelled graphs.

Example 8.36

We count the number of bipartite graphs, $\Gamma(V,E)$ with $|V| = 5$ and $V = U \cup W$ where $|U| = 3$ and $|W| = 2$. We use Polya's Theorem 6.27. The automorphism group is isomorphic to $\text{Sym}(3) \times \text{Sym}(2)$. If we represent the vertices of U by $\{1,2,3\}$ and those of W by $\{a,b\}$ we can represent the elements and their cycle types as in Table 8.3.

Element	Cycle type	# of this type
1	1^6	1
(ab)	2^3	1
(23)	$1^2 2^2$	3
(123)	3^2	2
$(ab)(23)$	2^3	3

Table 8.3 Cycles for bipartite graphs on five vertices.

We use B to represent an edge and W to represent no edge. From the table it is easy to find the cycle index polynomial.

$$\frac{1}{12} \left((B+W)^6 + (B^2+W^2)^3 + 3(B+W)^2(B^2+W^2)^2 + 2(B^3+W^3)^2 \right)$$
$$+ \frac{1}{12} \left(3(B^2+W^2)^3 + 2(B^6+W^6) \right).$$

By putting $B = W = 1$ we see there are 13 possible graphs of this type. If we expand in powers of B then we get

$$B^6 + B^5 W + 3B^4 W^2 + 3B^3 W^3 + 3B * 2W^4 + BW^5 + W^6.$$

So we can read from this that there are three bipartite graphs with three edges on graphs with bipartition of sizes 3 and 2.

8.3.3 Exercises

Exercise 8.11

Show that if a tree T with r vertices has exactly two vertices of degree 1 every other vertex has degree 2. Show that such trees are isomorphic but that such a tree can be labelled in $r!/2$ different ways.

Exercise 8.12

Let Γ be a connected graph. Prove that if each vertex of Γ has degree 2 then Γ is a cycle. Find the exponential generating function for the number of labelled graphs all of whose vertices have degree 2.

Exercise 8.13

Whilst there are three labelled trees on three vertices, how many unlabelled trees are there on three vertices? What happens for four vertices?

Exercise 8.14

Let Γ be a graph whose vertex set is $\{1,2,3,4\}$. How many labelled forests with two connected components are there?

Exercise 8.15

How many distinct graphs are there with five vertices?

8.4 Planarity

A very interesting class of graphs is those that can be drawn in the plane. Not all graphs can be drawn in the plane: for example the complete graph on five vertices cannot. There are number of ways to prove this, although playing around with pencil and paper will soon convince you. But we should be more precise about what we mean.

Definition 8.37

A graph is called a **plane graph** if it can be drawn in the plane so that the edges are simple curves which intersect only at its endpoints. A graph is called **planar** if it is isomorphic to a plane graph.

Obviously there is a topological aspect to this idea and we can think of a planar graph as one that can be embedded into the plane. Given a plane graph it divides the plane into simply connected regions called **faces**. Each face has a boundary of edges and we get the famous theorem:

Theorem 8.38 (**Euler's Theorem**)

Let Γ be a connected plane graph with v vertices, e edges and f faces. Then

$$v - e + f = 2.$$

Proof

The proof will be by induction on the number of edges. If there are no edges there is just one vertex and the result is certainly true. So now assume the result is true for a connected graph with n edges and assume that Γ is a connected planar graph with $n + 1$ edges.

If we choose an edge e so that $\Gamma - e$ is disconnected then for each connected component we have $v_1 - e_1 + f_1 = 2$ and $v_2 - e_2 + f_2 = 2$, where the two components are given by Γ_i with v_i vertices, e_i edges and f_i faces, $i = 1, 2$. But then $v_1 + v_2 = v$, $e_1 + e_2 = e - 1$ and $f_1 + f_2 = f + 1$, since we count two unbounded faces. So $v - e + f = 2$ as required.

Now if we can choose an edge so that the graph $\Gamma \setminus e$ is connected and e is on the dividing boundary between two faces then $\Gamma \setminus e$ has v vertices, $e - 1$ edges and $f - 1$ faces and so we are finished in this case. What happens if, when we delete e, there are the same number of faces? This can only happen if removing e disconnects the graph. This is a geometric argument: e lies on the boundary of only one face, say F. Then we can draw a curve from one side of e to the other within F. But now when we remove e the curve divides F into two and so disconnects the end points of e. $\qquad\square$

There are a number of geometric assumptions built into this proof, but a discussion of such issues would take us too far from our topic: for an interesting discussion see the book by Imre Lakatos [11].

Corollary 8.39

Let Γ be a connected plane graph with v vertices and e edges. Then $e \leq 3v - 6$. If, however, each face has at least four edges in its boundary then $e \leq 2v - 4$.

Proof

By the previous theorem $v - e + f = 2$, where f is the number of faces. Each face has at least three edges in its boundary, so $3f \leq 2e$, since each edge lies on the boundary of at most two faces. Now substitute in the above $3f = 6 - 3v + 3e \leq 2e$, which gives the corollary.

The second part follows similarly. $\square$

Example 8.40

The complete graph on five vertices has 10 edges. But $10 > 15 - 6$, so the complete graph on five vertices is not planar.

Example 8.41

The complete bipartite graph on six vertices has nine edges. But $9 < 18 - 6$, so this criterion is not sufficient to exclude this graph from being planar. However, $9 > 12 - 4$, so the second condition shows that the complete bipartite graph is not planar.

8.4.1 Exercises

Exercise 8.16

The girth g of a graph is defined to be the length of the shortest cycle in a graph. Let Γ be a graph with v vertices and e edges. Show that if Γ is planar then

$$(g - 2)e \leq g(v - 2).$$

Exercise 8.17

Show that every planar graph has a vertex of degree at most 5.

Exercise 8.18

Let Γ be a plane graph with less that 12 faces and every vertex has degree at least 3. Show that there is a face with no more than four edges in the boundary.

Exercise 8.19

Prove that if a graph Γ has more than 11 vertices then both Γ and its complement cannot be planar.

Exercise 8.20

A simple graph is called maximal planar if it is planar but adding any edge (on the given vertex set) would destroy that property. Show that all faces (even the outer one) are then bounded by three edges. Show that in this case that if the graph has v vertices with $v > 2$, then it has precisely $3v - 6$ edges and $2v - 4$ faces.

8.5 Miscellaneous Exercises

Exercise 8.21

Let $\Gamma = (V, E)$ be a graph on six vertices. Show that either Γ or its complement has a triangle as a subgraph.

Exercise 8.22

Let $\Gamma = (V, E)$ be a connected graph. The diameter d of a graph is the maximum distance apart of two vertices. If the maximal degree of any vertex is m prove that the number of vertices is bounded by

$$1 + m \frac{(m-1)^d - 1}{m-2}.$$

Exercise 8.23

Given a set of n men and a set of n women, let each man make a list of the women he is willing to marry. Then each man can be married to a woman on his list if, and only if, for every value of $k(1 \le k \le n)$ the union of any k of the lists contain at least k names.

Exercise 8.24

A group of friends buy a pizza and argue about how many pieces it can be cut into with a given number of cuts. Given that the pizza is circular and they use n cuts, find a formula for the maximum number of pieces.

Exercise 8.25

How many bipartite graphs with bipartitions of size 2 and 4 are there (see Example 8.36)?

9

Partitions and Paths

9.1 Introducing Partitions

The shopkeeper's problem – giving change to a customer who offers an amount of money for a particular purchase – leads to a very sophisticated mathematical idea.' Given one positive integer (the difference between the amount offered and the cost of the item) how many ways are there to write this (give change) as a sum of other positive integers?

Definition 9.1 (Partition)

A partition of a positive integer r by a set of positive integers A is a sum in which,

$$r = \sum_{a_i \in A} \lambda_i a_i$$

and each λ_i is a positive integer.

Giving change has no natural order, so the partitions of interest in this chapter are the **unordered** partitions (see Exercise 2.7, which deals with ordered partitions). Whenever the word partition is used in this chapter, we mean an **unordered partition**. As always, our interest will be to enumerative: in this case, to count the number of unordered partitions of an arbitrary positive integer, making use of a specified coinage.

A. Camina, B. Lewis, *An Introduction to Enumeration*,
Springer Undergraduate Mathematics Series,
DOI 10.1007/978-0-85729-600-9_9, © Springer-Verlag London Limited 2011

Example 9.2

We find the number of ways there are to partition (without regard to order) a given positive integer r using the positive integers 1 and 2. A generating function does the enumeration for us: we seek the coefficient of z^r in the product shown in Figure 9.1.

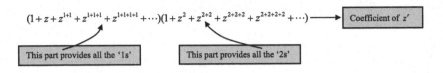

$$(1+z+z^{1+1}+z^{1+1+1}+z^{1+1+1+1}+\cdots)(1+z^2+z^{2+2}+z^{2+2+2}+z^{2+2+2+2}+\cdots) \longrightarrow \boxed{\text{Coefficient of } z^r}$$

<u>This part provides all the '1s'</u>   <u>This part provides all the '2s'</u>

Figure 9.1 Partitions using $\{1,2\}$.

If we begin to multiply out this product, we can find the number of such partitions for initial values of r:

$$1+z+2z^2+2z^3+3z^4+3z^5+4z^6+4z^7+\cdots.$$

A little thought should convince you that in order to be sure of the coefficient of z^r we must at least go as far as this power in each bracket.

In fact we can sidestep such a tedious expansion altogether.

Example 9.3

The expressions above may be written $(1+z+z^2+z^3+\cdots)$ and $(1+z^2+z^4+z^6+\cdots)$. We can rewrite this as

$$(1+z+z^2+z^3+\cdots)(1+z^2+z^4+z^6+\cdots) = \frac{1}{(1-z)(1-z^2)}.$$

Next we decompose this using partial fractions; it becomes

$$(1+z+z^2+z^3+\cdots)(1+z^2+z^4+z^6+\cdots) = \tfrac{1}{4}\left(\frac{2}{(1-z)^2}+\frac{1}{1-z}+\frac{1}{1+z}\right).$$

It is now an easy matter to pick out the required coefficient of z^r as

$$\tfrac{1}{4}(2(r+1)+1+(-1)^r) = \frac{2r+3+(-1)^r}{4}$$

and this is the required (explicit) answer. In fact we may write the answer in a more compact form, using the floor function

$$\text{Number of such partitions} = \left\lfloor \tfrac{r}{2} \right\rfloor + 1.$$

We can use the same basic ideas on a slightly more sophisticated example.

Example 9.4

A country has just three coins: a 1 cent coin, a 2 cent coin and a 4 cent coin – this is a very mathematical country. We seek the number of ways that change may be given for 100 cents. We do this by finding the coefficient of z^{100} in the generating function

$$(1+z+z^{1+1}+z^{1+1+1}\cdots)(1+z^2+z^{2+2}+z^{2+2+2}\cdots)(1+z^4+z^{4+4}+z^{4+4+4}\cdots)$$
$$=(1+z+z^2+\cdots)(1+z^2+z^4+\cdots)(1+z^4+z^8+\cdots)$$

Each of the bracketed expressions is an infinite GP and so for $|z| < 1$ we may rewrite it

$$=\frac{1}{(1-z)(1-z^2)(1-z^4)}.$$

Now we use partial fractions and find that

$$=\frac{1}{8(1-z)^3}+\frac{1}{4(1-z)^2}+\frac{9}{32(1-z)}+\frac{1}{16(1+z)^2}+\frac{5}{32(1+z)}+\frac{1+z}{8(1+z^2)}.$$

We may expand each term on the right using the Binomial Theorem; the coefficient of z^{100} is

$$\frac{1}{8}\binom{102}{2}+\frac{1}{4}\binom{101}{1}+\frac{9}{32}+\frac{(-1)^{100}}{16}\binom{101}{1}+\frac{5(-1)^{100}}{32}+\frac{(-1)^{50}}{8}=676.$$

Time to introduce some notation.

Definition 9.5 (Notation for partitions)

Given a set A of positive integers, the number of partitions of the positive integer r is

(i) written $p(r:A)$;

(ii) the "summands" in a partition are called **parts**.

Now we extend the results of Example 9.4 by seeking the number of ways to give change for an arbitrary amount, using the same coinage.

Example 9.6

The number of ways we can give change for an arbitrary amount (r cents) is simply $p(r:\{1,2,4\})$. But we already know the generating function for this:

$$\sum_{r\geq0} p(r:\{1,2,4\})z^r = \frac{1}{(1-z)(1-z^2)(1-z^4)}.$$

Again, we use partial fractions and as before

$$\sum_{r\geq0} p(r:\{1,2,4\})z^r = \frac{1}{8(1-z)^3} + \frac{1}{4(1-z)^2} + \frac{9}{32(1-z)}$$

$$+ \frac{1}{16(1+z)^2} + \frac{5}{32(1+z)} + \frac{1+z}{8(1+z^2)}.$$

We may expand each term on the right; the coefficient of z^r is complicated by the last three terms. We start with the final term since this determines how we deal with the other two:

$$\frac{1+z}{8(1+z^2)} = \frac{1}{8}(1+z)(1+z^2)^{-1} = \tfrac{1}{8}\left(1+z-z^2-z^3+z^4+\cdots\right)$$

so the coefficient of z^r from this term is

$$\begin{cases} \tfrac{1}{8} & \text{if } r=0,1(\bmod 4); \\ -\tfrac{1}{8} & \text{if } r=2,3(\bmod 4). \end{cases}$$

This tells us that we need to consider the remaining terms also in terms of residue classes mod 4:

$$\frac{1}{16(1+z)^2} = \frac{1}{16}\sum_{r\geq0}(-1)^r\binom{r+1}{1}z^r$$

so the coefficient of z^r from this term is

$$\begin{cases} \frac{r+1}{16} & \text{if } r=0,2(\bmod 4); \\ -\frac{r+1}{16} & \text{if } r=1,3(\bmod 4). \end{cases}$$

Finally

$$\frac{5}{32(1+z)} = \frac{5}{32}\sum_{r\geq0}(-1)^r z^r$$

so the coefficient of z^r from this term is

$$\begin{cases} \frac{5}{32} & \text{if } r=0,2(\bmod 4); \\ -\frac{5}{32} & \text{if } r=1,3(\bmod 4). \end{cases}$$

We now put all this together: when $r = 0(\bmod 4)$ the coefficient of z^r is,

$$
\begin{aligned}
p(r:\{1,2,4\}) &= \tfrac{1}{8}\binom{r+2}{2} + \tfrac{1}{4}\binom{r+1}{1} + \tfrac{9}{32} + \tfrac{r+1}{16} + \tfrac{5}{32} + \tfrac{1}{8} \\
&= \frac{r^2+8r+16}{16} \\
&= \frac{(r+1)(r+7)}{16} + \frac{9}{16} \\
&= \left\lceil \frac{(r+1)(r+7)}{16} \right\rceil
\end{aligned}
$$

where $\lceil\ \rceil$ is the 'ceiling' function (Definition 1.15). Looking at the four distinct possible values of $r\ (\bmod 4)$ we find that the number of ways of changing r cents is:

$$
\begin{cases}
\left\lceil \frac{(r+7)(r+1)}{16} \right\rceil & \text{for } r \equiv 0,2(\bmod 4) \\
\left\lceil \frac{(r+5)(r+1)}{16} \right\rceil & \text{for } r \equiv 1,3(\bmod 4).
\end{cases}
$$

Partitions come in all sorts of varieties so we will need additional notation to represent them.

Definition 9.7 (Types of partition)

(i) The number of partitions of r with parts taken from the natural numbers $\mathbb{N}$ is written in the form $p(r)$. Such partitions are called **unrestricted partitions**.

(ii) The number of partitions of r with parts taken from the set A (including the convention above – no set specified means the natural numbers) subject to a given condition on the parts, is written $p(r:A,\text{condition})$.

9.1.1 Unrestricted Partitions

These partitions are written $p(r)$ and have a generating function that is another infinite product, this time of infinite series:

$$
\sum_{r \geq 0} p(r)z^r = \left(1+z^1+z^{1+1}+\cdots\right)\left(1+z^2+z^{2+2}+\cdots\right)\left(1+z^3+z^{3+3}+\cdots\right)\cdots.
$$

Example 9.8

We may again expand this for the initial terms of the unrestricted partitions. We have

$$
\begin{aligned}
&(1+z+z^2+\cdots)(1+z^2+\cdots)(1+z^3+\cdots)(1+z^4+\cdots)\cdots \\
&= 1+z+2z^2+3z^3+5z^4+\cdots
\end{aligned}
$$

and so we have

$$\{p(r)\} = \{1,1,2,3,5,\ldots\}.$$

Again, we can simplify the generating function using sums of GPs.

Example 9.9

Each of the brackets in the generating function is an infinite GP, so when $|z| < 1$, we have

$$\sum_{r \geq 0} p(r)z^r = \frac{1}{(1-z)(1-z^2)(1-z^3)\cdots} = \prod_{k \geq 1} \frac{1}{(1-z^k)}.$$

We expand it as a power series and it is convergent whenever $|z| < 1$.

9.1.2 Distinct Partitions

Definition 9.10

A distinct partition consists entirely of parts that are distinct.

If the parts are to be distinct, no summand can be repeated. So the generating function for the distinct partitions is in the form of an infinite product of finite components, as in Figure 9.2.

$$\sum_{r \geq 0} p(r : \text{distinct parts})z^r = (1+z)(1+z^2)(1+z^3)\cdots = \prod_{k \geq 1} (1+z^k).$$

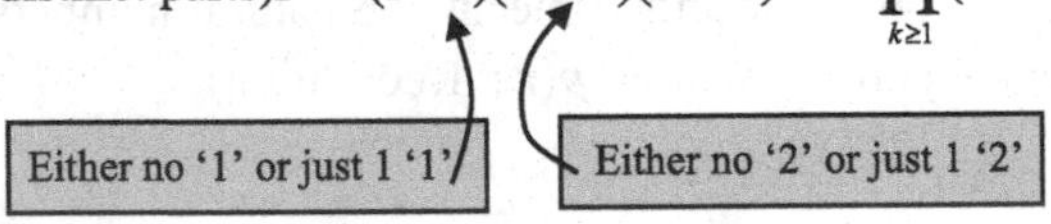

| Either no '1' or just 1 '1' | | Either no '2' or just 1 '2' |

Figure 9.2 Distinct partitions.

Example 9.11

If we begin to multiply out this product, we may find the initial terms of the sequence $\{p(r : \text{distinct parts}\}$:

$$\sum_{r \geq 0} p(r : \text{distinct})z^r = (1+z)(1+z^2)(1+z^3)(1+z^4)\cdots$$

$$= 1 + z + z^2 + 2z^3 + 2z^4 + \cdots.$$

This gives

$$\{p(r:\text{distinct})\} = \{1,1,1,2,2,\ldots\}$$

9.1.3 Odd Partitions

Definition 9.12

An odd partition consists of parts that are each odd positive integers.

Example 9.13

The generating function for these partitions is

$$\sum_{r\geqslant 0} p(r:\text{odd})z^r = (1+z^1+z^{1+1}+\cdots)(1+z^3+z^{3+3}+\cdots)\cdots$$

$$= \prod_{k\geqslant 0} \frac{1}{(1-z^{2k+1})}$$

The first few coefficients in the expansion give

$$\{p(r:\text{odd})\} = \{1,1,1,2,2,3,\ldots\}.$$

9.1.4 Relations Between Different Partitions

The generating functions are in a form that sometimes leads to astounding relations between apparently unconnected partitions. We start with just such a result.

Theorem 9.14 (Number of odd and distinct partitions are equal)

The number of partitions with odd parts is the same as those whose parts are distinct.

Proof (Using generating functions)

We have

$$\sum_{r\geq 0} p(r:\text{odd})z^r = \prod_{k\geq 1, k\text{ odd}} \frac{1}{(1-z^k)} = \frac{1}{(1-z)}\frac{1}{(1-z^3)}\cdots$$

$$= \frac{1}{(1-z)} \frac{1-z^2}{(1-z^2)} \frac{1}{(1-z^3)} \frac{1-z^4}{(1-z^4)} \cdots$$

$$= \frac{(1-z^2)}{(1-z)} \frac{(1-z^4)}{(1-z^2)} \frac{(1-z^6)}{(1-z^3)} \cdots = (1+z)\left(1+z^2\right)\left(1+z^3\right)\cdots$$

$$= \sum_{r\geq 0} p(r:\text{distinct})z^r.$$

$\square$

Many partition identities are discovered through generating functions, but there is a certain satisfaction in seeing an enumerative proof of the same result. It often provides better insight into how the particular partitions work. So now we give another (enumerative) proof of the same result.

Proof (Using an enumerative argument)

We use two constructions: the first (Figure 9.3) reduces an odd partition to a unique distinct partition. The second (Figure 9.4) works in reverse. They each exploit the fact that a positive integer can always be written uniquely as the product of a power of 2 and an odd number.

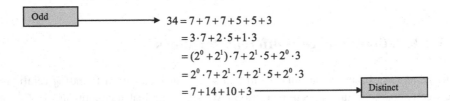

Figure 9.3 Odd partition to distinct partition.

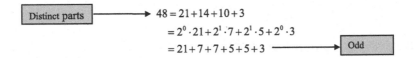

Figure 9.4 Distinct partition to odd partition.

The construction creates a bijection between the two types of partition and hence

there is an equal number of each. □

One more result.

Theorem 9.15

The number of partitions into parts not divisible by 3 is equal to the number of partitions in which each part appears at most twice.

Proof

The generating function of the partitions whose parts are not divisible by 3 is:

$$\sum_{r\geq 0} p(r : \text{not mult. of 3})z^r = \prod_{\substack{k\geq 1 \\ k\neq 3l}} \frac{1}{(1-z^k)} = \frac{1}{(1-z)}\frac{1}{(1-z^2)}\frac{1}{(1-z^4)}\cdots$$

$$= \frac{(1-z^3)}{(1-z)}\frac{(1-z^6)}{(1-z^2)}\frac{(1-z^9)}{(1-z^3)}\cdots.$$

But now we can divide out each "numerator" by the "denominator" below it:

$$\sum_{r\geq 0} p(r : \text{not mult. of 3})z^r = \left(1+z+z^2\right)\left(1+z^2+z^4\right)\left(1+z^3+z^6\right)\cdots$$

$$= \sum_{r\geq 0} p(r : \text{parts used at most, twice})z^r.$$

□

9.1.5 Ferrers Diagram

At the heart of many enumerative proofs of partition identities is a diagram that is a geometric representation of a partition.

Definition 9.16 (Ferrers diagram)

Any partition may be represented by a set of dots, called a Ferrers diagram:

(i) in which each part of the partition is represented by a proportionate, horizontal row of dots;

(ii) the diagram starts (at the top) with the largest part and proceeds downwards, part by part.

Example 9.17

The Ferrers diagram for the partition $23 = 7 + 6 + 4 + 4 + 2$ is the diagram:

Using this geometric idea, it is possible to give an immediate and intuitive proof of another partition identity.

Theorem 9.18

The number of partitions into exactly m parts is equal to the number of partitions whose largest part is m.

Proof

Any partition whose largest part is m has a Ferrers diagram like the representation on the left of Figure 9.5.

Figure 9.5 Ferrers diagram under diagonal reflection.

If we reflect it through the diagonal shown, it creates another partition whose Ferrers diagram is shown on the right. This has precisely m parts.   □

9.1.6 Partition Recurrences

We conclude with a recurrence for the unrestricted partitions first discovered by the mathematician Euler – his so-called Pentagonal Theorem. We start with the unrestricted partition generating function, and seek to derive a recurrence relation from the denominator of it. We have

$$\sum_{r\geq 0} p(r)z^r = \prod_{k\geq 1} \frac{1}{(1-z^k)}$$
$$\Rightarrow \prod_{k\geq 1}(1-z^k)\sum_{r\geq 0} p(r)z^r = 1.$$

At this point, Euler began expanding the expression that would make the required recurrence relation,

$$\prod_{k\geq 1}(1-z^k) = (1-z)\left(1-z^2\right)\left(1-z^3\right)\cdots$$

$$= 1-z-z^2+z^5+z^7-z^{12}-z^{15}+z^{22}+z^{26}-z^{35}-\cdots.$$

And this is where Euler made a decisive discovery. Not only were the coefficients $0, 1$ or -1, but the powers of z with non-zero coefficients come in pairs that on each occurrence are one further apart, with the first power in each such pair being a term of a sequence that Euler knew only too well: $\{1, 5, 12, 22, 35, \ldots\}$ – the Pentagonal sequence of numbers $\left\{\frac{r(3r-1)}{2}\right\}$, so called because they can be arranged as dot patterns of pentagons. So Euler believed that

$$\left(1+\sum_{r\geq 1}(-1)^r\left[z^{\frac{r(3r-1)}{2}}+z^{\frac{r(3r+1)}{2}}\right]\right)\left(\sum_{r\geq 0}p(r)z^r\right) = 1$$

and with this result, he could derive a powerful recurrence relation to calculate successive partition numbers.

Theorem 9.19 (Euler's Pentagonal Recurrence)

This states that

$$p(r) = \sum_{k\geq 1}(-1)^{k-1}\left(p(r-\tfrac{k(3k-1)}{2})+p(r-\tfrac{k(3k+1)}{2})\right) \text{ for } r\geq 1$$

that is

$$p(r) = p(r-1)+p(r-2)-p(r-5)-p(r-7)+\cdots.$$

We shall not prove this crucial result. Using it we find that

$$p(6) = p(5)+p(4)-p(1) = 7+5-1 = 11;$$
$$p(7) = p(6)+p(5)-p(2)-p(0) = 11+7-2-1 = 15;$$
$$\cdots$$
$$p(10) = p(9)+p(8)-p(5)-p(3) = 30+22-7-3 = 42;$$
$$p(11) = p(10)+p(9)-p(6)-p(4) = 42+30-11-5 = 56.$$

It is possible to continue in this way and use the recurrence to find values of $p(r)$ to any designated limit. For example, great perseverance gives

$$p(10) = 42$$
$$p(25) = 1,958$$
$$p(100) = 190,569,292$$
$$p(200) = 3,972,999,029,388.$$

For a discussion of this result and several proofs see [2, §14.4 & §14.5].

Clearly the partition function grows, but it does so in a very complicated fashion. There is another recurrence for the (unrestricted) partition numbers – this time we use calculus to establish it. Again there are two proofs – one that is based on generating functions and the other on enumeration. This uses a number theoretic function $\sigma_1(r)$, which returns the sum of the divisors of its argument.

Theorem 9.20 (The divisor/partition recurrence)

We have:

$$rp(r) = \sum_{k=1}^{r} \sigma_1(k) p(r-k).$$

Proof (Using generating functions)

We have

$$P(z) = \sum_{r \geq 0} p(r) z^r = \prod_{k \geq 1} \frac{1}{(1 - z^k)}$$

and taking logarithms of both sides gives

$$\log(P(z)) = \sum_{k \geq 1} \log(1 - z^k)^{-1} = \sum_{k \geq 1} \sum_{l \geq 1} \frac{z^{kl}}{l};$$

next we use the Euler operator (zD), which gives

$$\frac{z P'(z)}{P(z)} = \sum_{k \geq 1} \sum_{l \geq 1} k z^{kl} = \sum_{r \geq 1} \left(\sum_{\substack{m \geq 1 \\ m \text{ divides } r}} m \right) z^r$$

$$= \sum_{r \geq 1} \sigma_1(r) z^r$$

where $\sigma_1(r)$ is the sum of the divisors function. Rearranging this we have

$$zP'(z) = P(z) \sum_{r \geq 0} \sigma_1(r) z^r$$

and then comparing coefficients of z^r on either side gives the required recurrence. □

Example 9.21

We may use this recurrence to find $p(6)$. Using the recurrence, we have

$$6p(6) = \sigma_1(1)p_5 + \cdots + \sigma_1(6)p_0$$
$$= 66$$
$$\Rightarrow p(6) = 11.$$

This recurrence is not as efficient as Euler's Pentagonal recurrence; its significance is more that it relates partitions with the divisor function – two major functions of different branches of number theory.

Proof (Divisor/partition recurrence by enumeration)

First we add up all the partitions of r. There are $p(r)$ of them, so their sum is just

$$rp(r).$$

On the other hand, consider how many times the part n appears in these partitions. It appears at least once in $p(r-n)$ partitions; it appears at least twice in $p(r-2n)$ partitions; and so on. The total number of times it appears is $\sum_q p(r-qn)$. We must have

$$rp(r) = \sum_{n=1}^{r} n \sum_q p(r-qn) = \sum_{qn \leq r} np(r-qn) = \sum_{k=1}^{r} p(r-k) \sum_{q|k} q$$
$$= \sum_{k=1}^{r} p(r-k) \sigma_1(k)$$

exactly as required. □

9.1.7 Exercises

Exercise 9.1

Interpret the infinite product

$$P(z) = (1 + z + \cdots + z^9)(1 + z^{10} + \cdots + z^{90})(1 + z^{100} + \cdots + z^{900}) \cdots$$

as a partition generating function and so write down its expansion.

Exercise 9.2

Prove that the number of partitions in which each part appears at most four times is equal to the number of partitions into parts which are not divisible by 5.

Exercise 9.3

Find the generating function corresponding to the partitions in which:

(i) the largest part is 5;

(ii) the smallest part is 5;

(iii) each part is a power of 5.

Exercise 9.4

Prove that the number of partitions in which each part appears at least twice is equal to the number of partitions whose parts (mod 6) are not congruent to ± 1.

Exercise 9.5

If $p(r : \text{parts} \leq k)$ is the number of partitions in which the parts are no larger than k, establish the recurrence (using enumerative and generating function arguments)

$$p(r : \text{parts} \leq k) = p(r : \text{parts} \leq k-1) + p(r-k : \text{parts} \leq k).$$

Exercise 9.6

Consider the partitions whose parts are powers of 3, with no power used more than eight times. Denote the number of such partitions of r by $q(r)$. Show that they satisfy the recurrence $q(r) = q(r-1) + q(r-3) - q(r-4)$.

Exercise 9.7

Suppose that $p(r : k \text{ distinct parts})$ is the number of partitions in which there are k distinct parts. Using an enumerative argument, show that

$$p(r : k \text{ distinct parts}) = p(r-k : k-1 \text{ distinct parts}) + p(r-k : k \text{ distinct parts}).$$

Exercise 9.8

Show that $p(r : \text{odd parts})$ is

$$\sum_{k=0}^{r} p(k : \text{ distinct odd parts}) p(r-k : \text{even number of odd parts}).$$

Exercise 9.9

Show that

$$p(r : \text{distinct parts}) = p(r - 1 : \text{distinct parts}) + p(r : \text{distinct, not power of 2}).$$

Exercise 9.10

Determine $p(r : \{1, 2, 2^2, \ldots\})$.

9.2 Triangular Partitions of a Polygon

In 1751 Euler posed this problem to his fellow mathematician Goldbach: find the number of ways to partition a convex, plane polygon with r sides into triangles whose sides join vertices of the polygon and do not intersect. Goldbach, of course, may have been tied up with a number of competing conjectures; whatever, he seems not to have made any progress with the problem. But Euler did. He realized that if the polygon was convex, then the enumeration of these triangular partitions depended only on the number of sides of the polygon. We take this as a given. Euler had also found results for all such polygons with nine or fewer sides. He discovered that a triangle has only one such triangulation: itself. For a square it is almost as easy – there are just two ways to triangulate it. For a pentagon, things start to get complicated since there are five ways to triangulate it (Figure 9.6). Notice that some of these partitions are essentially

Figure 9.6 Triangular partitions of a pentagon.

the same if we are thinking geometrically – there are symmetries that map one into another, as in Chapter 6. However, here we are really dealing with a **labelled diagram** and each such partition is then distinct. Moving on to a hexagon in the same spirit, there are fourteen triangulations of a (regular) hexagon. Euler had soon developed the formula

$$\frac{2.6.10 \cdots (4r - 10)}{(r - 1)!}$$

for the number of such partitions, but was unable to prove it. This was no simple oversight on Euler's part. Despite its simple appearance, the problem is very difficult, and Euler's conjectured formula confirms that fact. The next step forward was made by another mathematician, Segner. Euler had sent on his first seven triangular partition numbers and Segner had responded with a recurrence relation. But one of a very strange and convoluted kind.

Definition 9.22 (Triangular partition numbers)

We call the number of triangular partitions of a convex polygon P_r with r sides, T_r.

Theorem 9.23 (Recurrence for triangular partitions)

The triangular partition numbers satisfy the recurrence

$$T_r = \begin{cases} 0 & : \quad r = 0, 1; \\ 1 & : \quad r = 2; \\ \sum_{k=0}^{r-1} T_k T_{r-k+1} & : \quad \text{otherwise.} \end{cases}$$

Proof

Consider a polygon with r sides, and one of its vertices – say the kth vertex, as shown in Figure 9.7.

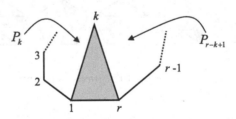

Figure 9.7 Triangular partitions of a convex polygon.

We form a triangle out of the three vertices, 1, k and r. To the left of this triangle is a (proper) convex, plane polygon with k vertices, denoted P_k; to the right is one with $r+1-k$ vertices, denoted P_{r-k+1}. So the number of distinct triangular partitions is

$$\#(P_k) . 1 . \#(P_{r-k+1}) = T_k T_{r-k+1}$$

But the vertex k can be chosen arbitrarily from $\{2,3,\ldots,r-1\}$ if we want $1, k, r$ to form a triangle as shown, and then:

$$T_r = \sum_{k=2}^{r-1} T_k T_{r-k+1} \quad \text{for } r \geq 4.$$

But we can simplify this; first we define $T_0 = T_1 = 0$ and also assign the value $T_2 = 1$. Then we have

$$T_r = \begin{cases} 0 & : \quad r = 0, 1 \\ 1 & : \quad r = 2; \\ \sum_{k=0}^{r-1} T_k T_{r-k+1} & : \quad \text{otherwise} \end{cases}$$

exactly as required. $\square$

We can convert this recurrence into a generating function.

Theorem 9.24 (Generating function for polygon partitions)

The generating function for the partitions of a convex polygon is

$$T(z) = \tfrac{z}{2}\left(1 - \sqrt{1-4z}\right).$$

Proof

The recurrence looks as if it comes from the product of two generating functions (*ogf*s) – and so it does. We have

$$T(z) = \sum_{r \geq 3} T_r z^r = T_2 z^2 + \sum_{r \geq 3} \left(\sum_{k=0}^{r-1} T_k T_r - k + 1 \right) z^r$$

$$= z^2 + \frac{1}{z} \sum_{r \geq 3} \left(\sum_{k=0}^{r-1} T_k T_{r+1-k} \right) z^{r+1}$$

and then using Theorem 5.17 we recognize this as

$$= z^2 + \frac{(T(z))^2}{z}.$$

Rearrangement of this gives

$$(T(z))^2 - zT(z) + z^3 = 0$$

which is a quadratic for the generating function $T(z)$. Solving it we find that

$$T(z) = \frac{z}{2}\left(1 - \sqrt{1-4z}\right).$$

$\square$

You will be asked to check in the exercises that the negative value of the square root is the correct choice. Expanding this generating function establishes Euler's formula.

Corollary 9.25

The explicit forms for the number of triangular partitions of a convex polygon of r sides (for $r \geq 3$) are

(i) $T_r = \frac{2^{r-2}1.3.\cdots(2r-5)}{(r-1)!}$

(ii) $T_r = \frac{1}{r-1}\binom{2r-4}{r-2}$.

You will be asked to prove these in the exercises.

The sequence of triangular partitions of a convex polygon $\{T_r\}$ has the initial terms

$$\{0,0,1,1,2,5,14,42,132,\ldots\}.$$

9.2.1 Exercises

Exercise 9.11

Given that $T_0 = T_1 = 0$ and $T_2 = 1$, use the recurrence of Theorem 9.23 to find the next three terms of the sequence $\{T_r\}$.

Exercise 9.12

The generating function $\frac{z}{2}\left(1 + \sqrt{1-4z}\right)$ is expanded. Find the first four terms in its power series expansion.

Exercise 9.13

By expanding the generating function of Theorem 9.24, establish the explicit forms given in Corollary 9.25.

9.3 Paths in a Lattice Plane

Consider a starting point (the origin $(0,0)$) in a lattice made up from the positive integer points. We will investigate the number of paths there are when we move according to the two rules shown in Figure 9.8.

$$(x,y) \mapsto (x+1,y) \quad \text{a unit horizontal move;} \quad \longrightarrow$$
$$(x,y) \mapsto (x,y+1) \quad \text{a unit vertical move.} \quad \uparrow$$

Figure 9.8 Allowable moves in a lattice.

Definition 9.26 (Number of paths: $\Gamma_{r,k}$)

(i) A path, according to the given set of rules, is called an **allowable** path.

(ii) We denote the number of allowable paths from the origin to the point (k,r) by $\Gamma_{r,k}$.

Note: the point (k,r) is in the $(r+1)$th row and in the $(k+1)$th column, so that the indexing is in reverse order to the coordinates in the plane.

Example 9.27

We can find the number of allowable paths to a defined point by exhausting the possibilities. The allowable paths from $(0,0) \rightarrow (2,2)$ are shown in Figure 9.9.

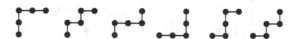

Figure 9.9 Paths from (0,0) to (2,2).

There are six such paths and hence $\Gamma_{2,2} = 6$.

Theorem 9.28 (A recurrence)

The number of allowable paths satisfies the recurrence

$$\Gamma_{r,k} = \Gamma_{r-1,k} + \Gamma_{r,k-1}.$$

Proof

All the paths to any given (positive) lattice point of the plane pass *either* through the

point below it, *or* the point to its left, as shown in Figure 9.10. This means that the
required recurrence is

$$\Gamma_{r,k} = \Gamma_{r-1,k} + \Gamma_{r,k-1}.$$

$\square$

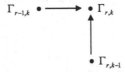

Figure 9.10 Paths from adjacent points.

Definition 9.29 (The path array)

The array $(\Gamma_{r,k} : r, k \geq 0)$ is called the path array of the allowable moves.

Using the recurrence it is easy to find the number of paths at each point by "creeping"
through the lattice to obtain the initial terms in the path array, as shown in Figure 9.11.

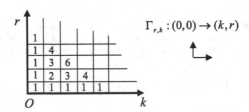

Figure 9.11 Initial terms in the array.

Given the initial conditions and the recurrence it follows that the number of paths
array has terms from Pascal's triangle. We have the following result.

Theorem 9.30 (Explicit form for the terms of the path array)

The number of allowable paths is given by

$$\Gamma_{r,k} = \binom{r+k}{k}.$$

There is an interesting enumerative argument that proves this result.

Proof

To move to the point (k, r) requires $r+k$ moves altogether; k of these must be horizontal and we immediately have $\Gamma_{r,k} = \binom{r+k}{k}$. Of course, we could equally have chosen to consider the number of vertical moves – it leads to the same result

$$\Gamma_{r,k} = \binom{r+k}{r} = \binom{r+k}{k}.$$

$\square$

9.3.1 Central Binomial Coefficients

If we consider the number of paths, $\Gamma_{r,r}$, there are to each (positive) lattice point on the main diagonal we find an interesting result. Using Theorem 9.30 we can give these terms an explicit form.

Corollary 9.31 (Paths to the main diagonal)

The number of allowable paths to the main diagonal is

$$\Gamma_{r,r} = \binom{r+r}{r} = \binom{2r}{r}.$$

Note: the terms here are the central (even rows) Binomial coefficients. Consideration of such paths leads to another enumerative identity.

Theorem 9.32

The central Binomial coefficients are given by:

$$\binom{2r}{r} = \sum_{k \geq 0} \binom{r}{k}^2.$$

Proof

The number of paths to (r,r) is given by $\binom{2r}{r}$. However, after r moves, suppose that the path has attained a "height" of k units (so that $0 \le k \le r$). So in the next r moves it must ascend a further $r - k$ units. We deduce that

$$\Gamma_{r,r} = \sum_{k=0}^{r} \binom{r}{k}\binom{r}{r-k} = \sum_{k \ge 0} \binom{r}{k}^2$$

and hence the result.   □

This was proved in a different enumerative way in Exercise 5.23.

9.3.2 Catalan Numbers

Now we change the nature of the allowable paths we consider.

Definition 9.33 (Paths that do not rise above the diagonal)

The allowable paths in this case are those paths that do not rise above the main diagonal. We denote the number of such paths by $\varphi_{r,k}$ and obviously, $k \le r$.

Again, it is easy to "creep" along the initial lattice points to obtain initial terms in the path array as shown in Figure 9.12.

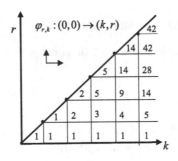

Figure 9.12 Paths that don't rise above the diagonal.

The number of such paths to the (positive) lattice points on the main diagonal is the sequence that starts

$$\{\varphi_{r,r}\} = (1, 1, 2, 5, 14, \ldots)$$

They appear, astonishingly, to be the Triangular partition numbers of the last section – slightly shifted. This celebrated sequence, called the **Catalan numbers**, is named after the mathematician who first investigated their properties. To show that these very different enumerative problems (triangular partitions and lattice paths not rising above the diagonal) are essentially the same, we first prove two preliminary results in the form of lemmas.

Lemma 9.34

The number of allowable paths from $(0,0)$ to (r,r) that do not touch the main diagonal until (r,r) is denoted by u_r. Then

$$u_r = \begin{cases} 1 & : \quad r = 0; \\ \varphi_{r-1,r-1} & : \quad \text{otherwise.} \end{cases}$$

Proof

Such a path (for $r \geq 1$) must start with a horizontal move and finish with a vertical move. In between, it can take any path in the shaded triangle (of Figure 9.13) that does not rise above the "hypotenuse" of the triangle.

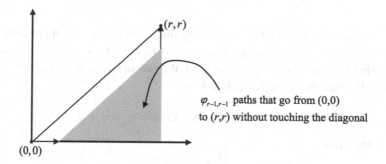

(r,r)

$\varphi_{r-1,r-1}$ paths that go from $(0,0)$
to (r,r) without touching the diagonal

$(0,0)$

Figure 9.13 Paths that don't meet the diagonal.

There are precisely $\varphi_{r-1,r-1}$ such paths inside the shaded triangle, as required. $\square$

Now we state and prove another result: in fact the terms of the Catalan sequence are just the triangle partition numbers suitably displaced. We do so by showing that they obey the same recurrence and share (essentially) the same initial terms.

Lemma 9.35

The Catalan sequence $(\varphi_{r,r})$ that counts the number of paths that do not meet the main diagonal before ending on it is related to the triangular partitions of a convex polygon (T_r) by

$$\varphi_{r,r} = T_{r+2}.$$

Proof

Any path that goes from $(0,0)$ to (r,r) and does not rise above the main diagonal meets one of the points $(0,0)$ *or* $(1,1)$ *or* ... *or* $(r-1,r-1)$ *or* (r,r) first. Suppose that it meets (first) the point (k,k) on the diagonal, as shown in Figure 9.14.

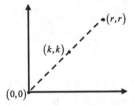

Figure 9.14 First point it meets on the diagonal.

The number of paths from $(0,0)$ to (k,k) that do not touch the main diagonal is (by Lemma 9.34) $\varphi_{k-1,k-1}$. The number of paths from (k,k) to (r,r) that do not rise above the main diagonal is $\varphi_{r-k,r-k}$. So in all, the number of paths from $(0,0)$ to (r,r) that touch the main diagonal (first) at (k,k) is

$$\varphi_{k-1,k-1}\varphi_{r-k,r-k}.$$

It now follows that

$$\varphi_{r,r} = \sum_{k=1}^{r} \varphi_{k-1,k-1}\varphi_{r-k,r-k}$$

and this is the same recurrence as that for the triangular partitions of a convex polygon with the initial conditions that $\varphi_{0,0} = \varphi_{1,1} = 1$. □

With this relation between the triangular partitions $\{T_r\}$ and the Catalan numbers $\{\varphi_{r,r}\}$ we can derive other properties of the latter.

Theorem 9.36

We have:

(i) the explicit form for the terms of the Catalan sequence is

$$\varphi_{r,r} = \frac{1}{r+1}\binom{2r}{r}$$

(ii) its generating function is

$$\sum_{r\geq 0}\phi_{r,r}z^r = \frac{1-\sqrt{1-4z}}{2z}.$$

Proof

We have from Lemma 9.35 and Corollary 9.25

$$\varphi_{r,r} = T_{r+2} = \frac{1}{r+2-1}\binom{2(r+2-2)}{r+2-2} = \frac{1}{r+1}\binom{2r}{r}.$$

Then:

$$\sum_{r\geq 0}\varphi_{r,r}z^r = \sum_{r\geq 0}T_{r+2}z^r = \frac{1}{z^2}\sum_{r\geq 0}T_{r+2}z^{r+2} = \frac{1}{z^2}\sum_{r\geq 0}T_r z^r$$

and hence

$$\sum_{r\geq 0}\phi_{r,r}z^r = \frac{1-\sqrt{1-4z}}{2z}.$$

$\square$

The Catalan sequence is usually written $\{C_r\}$ and has the initial terms

$$\{C_r\} = \{1,1,2,5,14,\ldots\}.$$

9.3.3 Exercises

Exercise 9.14

Catalan first discovered the sequence of numbers that bear his name, $\{C_r\}$, when he investigated the problem of determining the number of ways that a list of $r+1$ terms $\{\alpha_0, \alpha_1, \ldots, \alpha_r\}$ could be bracketed so that the order of multiplication is completely specified. For example, the number C_2 is the number of ways to bracket $\{\alpha_0, \alpha_1, \alpha_2\}$ and it is $\#\{(\alpha_0)(\alpha_1, \alpha_2), (\alpha_0, \alpha_1)(\alpha_2)\} = 2$. We also have $C_3 = 5$. Construct a recurrence for this sequence.

Exercise 9.15

By simplifying the expression $\binom{2r}{r} - \binom{2r}{r+1}$ show that the Catalan number formula, $C_r = \frac{1}{r+1}\binom{2r}{r}$, does yield integers.

Exercise 9.16

Find the generating function of the Catalan numbers from the generating function

$$\sum_{r\geq 0}\binom{2r}{r}z^r = \frac{1}{\sqrt{1-4z}}.$$

Exercise 9.17

Show that the Catalan numbers have the nice summation property that

$$\sum_{r\geq 0}\frac{C_r}{4^r} = 2.$$

9.4 Miscellaneous Exercises

Exercise 9.18

Show that $\left(1+z+z^2\right)\left(1+z^3+z^6\right)\left(1+z^9+z^{18}\right)\cdots = \frac{1}{1-z}$.

Exercise 9.19

Consider the number of partitions of r, $q(r)$, whose parts are powers of 3, with no power used more than eight times. You were invited to establish a recurrence in Exercise 9.6. Now use partial fractions to derive an explicit form for $q(r)$.

Exercise 9.20

Find an explicit form for the number of partitions of r which consist of powers of 2, where no power can be used more than three times.

Exercise 9.21

Show that the number of partitions with no parts equal to 3, $q(r)$, is related to unrestricted partitions by $q(r) + p(r-3) = p(r)$.

Exercise 9.22

The divisor/partition recurrence (Theorem 9.20)

$$rp(r) = \sum_{k=1}^{r}\sigma_1(k)p(r-k)$$

may be written in matrix form. Use the initial terms of this matrix, show that

$$
\begin{pmatrix} \sigma_1(1) \\ \sigma_1(2) \\ \sigma_1(3) \\ \sigma_1(4) \\ \cdot \end{pmatrix} = \begin{pmatrix} 1 & & & & \\ -1 & 1 & & & \\ -1 & -1 & 1 & & \\ 0 & -1 & -1 & 1 & \cdot \\ \cdot & \cdot & \cdot & \cdot & \cdot \end{pmatrix} \begin{pmatrix} 1p(1) \\ 2p(2) \\ 3p(3) \\ 4p(4) \\ \cdot \end{pmatrix} .
$$

Exercise 9.23

Show that the number of paths to (r,r) that rise above the diagonal is $\binom{2r}{r-1}$.

Exercise 9.24

How many "mountain ranges" can be formed with r upstrokes and r down-strokes? An example is shown in Figure 9.15.

$r=1$, 1 way $r=2$, 2 ways

Figure 9.15 "Mountain ranges".

Exercise 9.25

Prove that the Catalan numbers have the product form

$$
C_r = \frac{1}{(r+1)!} \prod_{k=1}^{r} (4k-2).
$$

Exercise 9.26

Derive the identity

$$
\sum_{k \geq 0} \binom{2k}{k} \binom{2r-2k}{r-2} = 4^r.
$$

A

Library

Ordinary Generating Functions (*ogf*)

The throw of r dice and the score obtained $\xrightarrow{ogf} (z + z^2 + \cdots z^6)^r$

$$\{1,1,1,\ldots\} \xrightarrow{ogf} \frac{1}{1-z}$$

$$\{0,1,1,\ldots\} \xrightarrow{ogf} \frac{z}{1-z}$$

$$\{1,2,3,\ldots\} \xrightarrow{ogf} \frac{1}{(1-z)^2}$$

The Fibonacci sequence

$$\{F_r\} = \{0,1,1,2,3,\ldots\} \xrightarrow{ogf} \frac{z}{1-z-z^2}$$

The Lucas sequence

$$\{L_r\} = \{2,1,3,4,7,\ldots\} \xrightarrow{ogf} \frac{2-z}{1-z-z^2}$$

The triangular partition sequence

$$\{T_r\} = \{0,0,1,1,2,5,\ldots\} \xrightarrow{ogf} \frac{z(1-\sqrt{1-4z})}{2}$$

The Central Binomial coefficients

$$\{\tbinom{2r}{r}\} = \{1,2,6,20,\ldots\} \xrightarrow{ogf} \frac{1}{\sqrt{1-4z}}$$

A. Camina, B. Lewis, *An Introduction to Enumeration*,
Springer Undergraduate Mathematics Series,
DOI 10.1007/978-0-85729-600-9, © Springer-Verlag London Limited 2011

The Catalan sequence

$$\{C_r\} = \{1,1,2,5,14,\ldots\} \xrightarrow{ogf} \frac{\left(1-\sqrt{1-4z}\right)}{2z}$$

Unrestricted partitions

$$\{p(r)\} = \{1,1,2,3,5,\ldots\} \xrightarrow{ogf} \frac{1}{(1-z)(1-z^2)(1-z^3)\cdots}$$

Partitions whose parts are distinct

$$\{p(r:\text{distinct})\} = \{1,1,1,2,2,3,\ldots\} \xrightarrow{ogf} (1+z)(1+z^2)(1+z^3)\cdots$$

Partitions whose parts are odd

$$\{p(r:\text{odd})\} = \{1,1,1,2,2,3,\ldots\} \xrightarrow{ogf} \frac{1}{(1-z)(1-z^3)(1-z^5)\cdots}$$

Exponential Generating Functions (*egf*)

$$\{u_r\} = \{1,1,1,\ldots\} \xrightarrow{egf} e^z$$

The Fibonacci sequence

$$\{F_r\} = \{0,1,1,2,3,\ldots\} \xrightarrow{egf} \frac{1}{\sqrt{5}}\left(e^{\phi_1 z} - e^{\phi_2 z}\right)$$

The Lucas sequence

$$\{L_r\} = \{2,1,3,4,7,\ldots\} \xrightarrow{egf} e^{\phi_1 z} + e^{\phi_2 z}$$

The Derangement sequence

$$\{d_r\} = \{1,0,1,2,9,\ldots\} \xrightarrow{egf} \frac{e^{-z}}{1-z}$$

The push-button lock sequence

$$\{c_r\} = \{1,2,5,16,65,\ldots\} \xrightarrow{egf} \frac{e^z}{1-z}$$

Involutions – permutations all of whose cycles are of length 2 or less

$$\{\sigma(r:\text{cycles} \le 2)\} = \{1,1,2,4,10,\ldots\} \xrightarrow{egf} e^{z+\frac{z^2}{2}}$$

Permutations all of whose cycles are of length 2

$$\{\sigma(r:\text{cycles} = 2)\} = \{1,0,1,0,3,0,\ldots\} \xrightarrow{egf} e^{\frac{z^2}{2}}$$

Permutations all of whose cycles are of length 3 or more

$$\{\sigma(r : \text{cycles } 2\} = \{1,0,0,2,6,\ldots\} \xrightarrow{egf} \frac{e^{-z-\frac{z^2}{2}}}{1-z}$$

The zig-zag permutations

$$\{a_r\} = \{1,1,1,2,5,16,\ldots\} \xrightarrow{egf} \tan z + \sec z$$

Solutions

Chapter 1

1.1 $\binom{r}{k}$ counts the number of ways to select k from r, which also counts selecting $r-k$ from r, since they are selected by not being selected.

1.2 We can choose one from each r, r and r in $\left(\binom{r}{1}\right)^3$ ways. We can choose two from one r, one from another r and none from the final r in $\binom{r}{2}\binom{r}{1}$ ways and there are six ways of doing this. We can choose three from one r and none from the others in $\binom{r}{3}$ ways. There are three possible ways to do this. The result follows.

1.3 Thirteen socks.

1.4 Seven times.

1.5 (i) Split up the unit equilateral triangle into four equilateral triangles with sides $\frac{1}{2}$. If five points are placed in the triangle, then at least two of them must be in one of the smaller triangles – and must be separated by less than $\frac{1}{2}$. (ii) Split up the unit equilateral triangle into nine equilateral triangles with sides $\frac{1}{3}$. So 10 points placed inside must be less that $\frac{1}{3}$ apart.

1.6 Of two consecutive integers, one of them is divisible by 2; of three consecutive integers, one of them is divisible by 3; etc.

1.7 The number is $6300 - (2100 + 1575 + 1260) + (525 + 420 + 315) - 105 = 2520$.

1.8 (i) 2 is fixed in 3! ways. So $4! - 3! = 18$ counts the number of ways that 2 is not fixed. (ii) 2 and 4 are fixed in 2! ways. So $4! - (3! + 3!) + 2! = 14$ counts the number of ways that 2 and 4 are not fixed.

1.9 The number of integers relatively prime to r is $r(1 - \frac{1}{p_1})(1 - \frac{1}{p_2})$.

1.10 Consider three cases when x is: an integer; has fractional part $< \frac{1}{2}$; has fractional

A. Camina, B. Lewis, *An Introduction to Enumeration*,
Springer Undergraduate Mathematics Series,
DOI 10.1007/978-0-85729-600-9, © Springer-Verlag London Limited 2011

part $\geq \frac{1}{2}$.

1.11 To choose k from $r+l$ we can choose m from r and $k-m$ from l for $0 \leq m \leq k$.

1.12 $24 = b_4 + 6.1 + 3.1$, which gives $b_4 = 15$; $5! = b_5 + \binom{5}{2}b_3 + \frac{1}{2!}\binom{5}{2}\binom{3}{2}b_1$ which gives $b_5 = 75$.

Chapter 2

2.1 The generating function is easily seen to be $(z + z^2 + \cdots + z^6)^{10}$, which becomes $z^{10}(1 - z^6)^{10}(1 - z)^{-10}$. We require the coefficient of z^{25}, which is

$$\binom{10}{0}\binom{24}{15} - \binom{10}{1}\binom{18}{9} + \binom{10}{2}\binom{12}{3}.$$

2.2 From Example 2.7 the answer is 676.

2.3 The generating function is $\sum_{r \geq 0} u_r z^r = \frac{1}{(1-z)(1-z^3)}$ and using partial fractions this is $\frac{1}{3}\left(\frac{1}{1-z} + \frac{1}{(1-z)^2} + \frac{1-z^2}{1-z^3}\right)$. So the required coefficient is $\frac{1}{3}(1 + 201 - 1) = 67$. An alternative approach is to consider how many 3p coins will be used, so the answer is $\lceil (200/3) \rceil = 67$.

2.4 1, p_1 and p_1^2 are the divisors of p_1^2 and 1, p_2, p_2^2 and p_2^3 are the divisors of p_2^3 and hence the result.

2.5 Consider a legal expression of length r. It must either end in a double digit or in an operation followed by a digit. In the first case we may add another digit to one of the a_{r-1} legal expressions in $10a_{r-1}$ ways. In the second case, from one of the a_{r-2} legal expressions we can add 29 combinations of an operation and a digit – excluding division by zero. So $a_r = 10a_{r-1} + 29a_{r-2}$.

2.6 A legal string of length r either starts with a 0 and must then be followed by a 1 and a legal string of length $r - 2$, or it starts with a 1 and can be followed by any legal string of length $r - 1$. Hence $u_r = u_{r-2} + u_{r-1}$ and the initial conditions then give $u_r = F_{r+2}$.

2.7 Consider the final summand: it is either 1 or 2. So $u_{r+1} = u_r + u_{r-1}$ and hence $u_r = F_{r+1}$.

2.8 We have

$$rd_{r-1} = r\sum_{k=0}^{r-1}(-1)^k\binom{r-1}{k}(r-k-1)! = \sum_{k=0}^{r-1}(-1)^k\binom{r}{k}(r-k)!$$

$$= \sum_{k=0}^{r}(-1)^k\binom{r}{k}(r-k)! - (-1)^r = d_r - (-1)^r$$

as required. We use this for consecutive values of r: $d_r = rd_{r-1} + (-1)^r$ and $d_{r-1} = (r-1)d_{r-2} + (-1)^{r-1}$ and addition gives $d_r + d_{r-1} = rd_{r-1} + (r-1)d_{r-2}$ so that $d_r = (r-1)(d_{r-1} + d_{r-2})$, as required.

2.9 By definition, $L_r = F_{r-1} + F_r + 1$ and so $L_r + L_{r+1} = F_{r-1} + F_{r+1} + F_r + F_{r+2} = 2F_{r+1} + F_{r+2} = F_{r+1} + F_{r+3} = L_{r+2}$, as required.

2.10 (i) $\frac{1}{1+z}$; (ii) $\frac{1}{(1+z)^2}$; (iii) $\frac{1}{2}(\frac{1}{(1+z)^2} + \frac{1}{(1-z)^2})$.

2.11 $\log\frac{1}{1-z}$.

2.12 $\frac{4+z}{1-z-2z^2}$.

2.13 The generating function for a_r is $\frac{4-4z}{1-2z-15z^2}$.

2.14 $L(z) - 2 - z = z(L(z) - 2) + z^2L(z)$ gives $L(z) = \frac{2-z}{1-z-z^2}$.

2.15 Either a string of length r has one or more 0s or it has no 0s. So $u_{r+1} = 3u_r + 3^r$. The generating function is $\frac{1-2z}{(1-3z)^2}$ (which gives $u_r = (r+3)3^{r-1}$).

2.16 It is easy to prove (by induction) that $M^r = \begin{pmatrix} F_{r-1} & F_r \\ F_r & F_{r+1} \end{pmatrix}$. The first result easily follows from the definition of Lucas numbers. We also have $|M| = -1$ and hence $|M^r| = (-1)^r$. But we can evaluate the determinant directly, and $|M^r| = F_{r-1}F_{r+1} - F_r^2$ and the result follows.

2.17 The generating function is $\frac{1}{(1-z)(1-z^2)} = \frac{1}{4}(\frac{1}{1-z} + \frac{1}{1+z} + \frac{2}{(1-z)^2})$. (So the required number is $\frac{2r+3+(-1)^r}{4}$).

2.18 In each module, the mark is generated by $(1+z+z^2+\cdots+z^m)$. So we require the coefficient of z^{3m} in $(1+z+z^2+\cdots+z^m)^6$. Rewrite this as $\frac{(1-z^{m+1})^6}{(1-z)^6} = (1-6z^{m+1} + 15z^{2m+2} + \cdots)\sum_{s\geq0}\binom{5+s}{5}z^s$. The required number follows.

2.19 A sum that makes r ends in a 1, a 2 or a 3. So $p_r = p_{r-1} + p_{r-2} + p_{r-3}$.

2.20 A tiling either ends with a configuration of tiles that takes up a single horizontal space (one double tile placed vertically, or two single tiles placed on top of each other), or with a configuration of tiles that takes up a double horizontal space (one horizontal double on top of two singles; or reverse this; or two doubles on top of each other). This gives the required recurrence. The generating function is $\frac{1}{1-2z-3z^2}$ with $f_0 = 1$ and $f_1 = 2$.

Chapter 3

3.1 $v_r = 4^r + 3.4^{r-1} = 7.4^{r-1}$.

3.2 For example, $\frac{(1+\sqrt{5})^3 - (1-\sqrt{5})^3}{2^3 \cdot \sqrt{5}} = 2 = F_3$.

3.3 If $(1 - 3z + z^2) = (1 - \theta_1 z)(1 - \theta_2 z)$ then $\theta_1 + \theta_2 = 3$ and $\theta_1 \theta_2 = 1$. This gives $\theta_{1,2} = \frac{3 \pm \sqrt{5}}{2}$. We find that $u_r = \frac{(3+\sqrt{5})^r + (3-\sqrt{5})^r}{2^r} = \frac{(1+\sqrt{5})^{2r} + (1-\sqrt{5})^{2r}}{2^{2r}} = L_{2r}$.

3.4 This time, $v_r = F_{2r+1}$.

3.5 $L_{2r} = \phi_1^{2r} + \phi_2^{2r}$ and so $\sum_{r \geq 0} L_{2r} z^r = \sum_{r \geq 0} (\phi_1^2 z)^r + \sum_{r \geq 0} (\phi_2^2 z)^r = \frac{1}{1 - \phi_1^2 z} + \frac{1}{1 - \phi_2^2 z}$

$= \frac{2 - (\phi_1^2 + \phi_2^2) z}{1 - (\phi_1^2 + \phi_2^2) z + \phi_1^2 \phi_2^2 z^2} = \frac{2 - 3z}{1 - 3z + z^2}$. The recurrence follows from the denominator. Similarly, $\sum_{r \geq 0} L_{3r} z^r = \frac{2 - 4z}{1 - 4z - z^2}$.

3.6 $\sum_{r \geq 1} u_r z^r = 2z \sum_{r \geq 1} u_{r-1} z^{r-1} + 2 \sum_{r \geq 1} (2z)^{r-1} + 3 \sum_{r \geq 1} (3z)^{r-1}$ and so $U(z) - 1 = 2zU(z) + \frac{2}{1-2z} + \frac{3}{1-3z}$ and so $U(z) = \frac{6 - 17z + 6z^2}{(1-2z)^2(1-3z)}$ which leads to the recurrence $u_r = 7u_{r-1} - 16u_{r-2} + 12u_{r-3}$.

3.7 The characteristic equation is $\theta^2 - 2\theta + 2 = 0$, which has roots $\theta = 1 \pm i$. This leads to $u_r = \left(\frac{1-2i}{2}\right)(1+i)^r + \left(\frac{1+2i}{2}\right)(1-i)^r$.

3.8 $n_{r+1} = 3.10^r + 4n_r$ and so $10n_r = 3.10^r + 40n_{r-1}$. Subtraction of these gives $n_{r+1} = 14n_r - 40n_{r-1}$. This leads to $n_r = C4^r + D10^r$, and the initial values determined in Example 2.11 then give $C = D = \frac{1}{2}$.

3.9 The sequence proceeds $\{3, 0, 3, 3, 6, 9, 15, \ldots\} = 3\{1, 0, 1, 1, 2, 3, 5, \ldots\}$. We conjecture that $G_r = 3F_{r-1}$. This is easily proved by induction.

3.10 We have $a_{r+1} + \sqrt{3}b_{r+1} = (1 + \sqrt{3})(a_r + \sqrt{3}b_r)$. Multiplying out the right-hand side of this and then comparing rational and irrational parts (the sequences a_r and b_r are each integer sequences) gives the required results. From the first of these we have $b_r = \frac{a_{r+1} - a_r}{3}$ and substituted into the second gives (when rearranged) $a_{r+2} = 2a_{r+1} + 2a_r$. The sequence $\{b_r\}$ satisfies the same recurrence. We then have $a_r = \frac{1}{2}((1 + \sqrt{3})^r + (1 - \sqrt{3})^r)$ and $b_r = \frac{1}{2\sqrt{3}}((1 + \sqrt{3})^r - (1 - \sqrt{3})^r)$.

3.11 The recurrence, and the initial values $f_0 = 1$ and $f_1 = 2$, leads to the generating function $F(z) = \frac{1}{1 - 2z - 3z^2}$. Using partial fractions and expanding then gives $f_r = \frac{3^{r+1} + (-1)^r}{4}$.

3.12 The generating function satisfies $\frac{1}{z}(A(z) - 1) = 3A(z) + \frac{2}{1-2z}$, which becomes (using partial fractions) $A(z) = \frac{3}{1-3z} - \frac{2}{1-2z}$ and hence $a_r = 3^{r+1} - 2^{r+1}$.

3.13 Suppose there are u_r permutations of $\{1, 2, \ldots, r\}$ that move each element fewer than two places. Now consider u_{r+1}. Where does $r+1$ go? It must *either* go in the penultimate place (in which case it must be followed by r) *or* it goes in the final place. So $u_{r+1} = u_r + u_{r-1}$. Using the initial conditions that $u_1 = 1$ and $u_2 = 2$ this means that $u_r = F_{r+1}$.

3.14 Either a word starts with an e (which must then be followed by an a) or it doesn't (so it starts with a or t). So $w_r = w_{r-2} + 2w_{r-1}$ and $w_0 = 1, w_1 = 2$. Using Theorem 3.23 this gives $w_r = \frac{1}{2\sqrt{2}}((1+\sqrt{2})^{r+1} - (1+\sqrt{2})^{r+1})$.

3.15 A word of length r ends in either v (so the next letter is e,r, or y); or e (so the next letter is v, r or y); or r (so the next letter is v, e or y); or y (so the next letter is v, e or r). It follows that $w_{r+1} = 3w_r$ and then $w_r = 4.3^{r-1}$ for $r \geq 1$.

3.16 We have (from Example 3.4) $\phi_1 + \phi_2 = 1$ and multiplication by ϕ_1 gives $\phi_1^2 - 1 = \phi_1$; hence the required result. (ϕ_1 and ϕ_2 are always interchangeable.) If we multiply the result again by ϕ_1 we find $\phi_1^3 = 2\phi_1 + 1$. Induction proves the remaining result.

3.17 (i) These polynomials follow readily from the given recurrence. (ii) As the indeterminate "w" is independent of r, we may construct the given characteristic equation as given. (iii) Solving the characteristic equation gives $\theta_{1,2} = \frac{w \pm \sqrt{w^2+4}}{2}$ and this leads to the given explicit form of the polynomials.

3.18 If the denominator and numerator of the generating function of the sequence $\{u_r\}$ are, respectively, $a_0 + a_1 z + \cdots + a_p z^p$ and $b_0 + b_1 z + \cdots + b_q z^q$ then $\left(\sum_{t=0}^{p} a_t z^t\right) \sum_{r \geq 0} u_r z^r = \sum_{s=0}^{q} b_s z^s$ and comparing coefficients of z^r when $r > q$ gives the recurrence $\sum_{t=0}^{p} a_t u_t = 0$.

3.19 The given expression is (using partial fractions) $1 + \frac{1}{(1-z)2} - \frac{1}{1-z}$ and then comparing coefficients gives $u_r - u_{r-6} = (r+1) - 1$ for $r \geq 7$.

Chapter 4

4.1 Let H_i be a family of subgroups of G indexed by a set I. Let $\cap_{i \in I} H_i = H$. Let $x, y \in H$. Then $x, y \in H_i$ for all $i \in I$. By the properties of subgroups we have that xy and x^{-1} are both in H_i for all $i \in I$, so both are in H. So H is a subgroup of G. Now for any subset X consider the family of subgroups which contain X. This is not empty as G is in the family. Define $\langle X \rangle$ to be the intersection of this family. It is now clear that $\langle X \rangle$ is the smallest subgroup of G containing X. This is called the subgroup generated by X and is written $\langle X \rangle$.

4.2 Consider the set $\{g^m : m \in \mathbb{Z}\}$. Note that the number of elements in this set is n. So amongst the set $\{g^0, g^1, \ldots, g^n\}$ there must be two elements the same. Say $g^i = g^j$ for

$0 \leq i, j \leq n$. Then $g^{j-i} = 1$. So there are positive natural numbers m so that $g^m = 1$. Let m be the smallest. We now claim that $\{g^0, g^1, \ldots, g^{m-1}\} = \langle g \rangle$ and hence that $m = n$ as claimed. Let $k \in \mathbb{Z}$. We can write $k = am + r, 0 \leq r < m$. Then

$$g^k = g^{am+r} = g^{am} g^r = (g^m)^a g^r = g^r.$$

So $g^k \in \{g^0, g^1, \ldots, g^{m-1}\}$.

Now let $g^d = 1$ and put $d = am + r, 0 \leq r < m$. As above $g^d = g^r = 1$ but by the previous paragraph we see that $r = 0$.

4.3 Let ϕ be a homomorphism from G to H.

(i) Let $K = \{g : g \in G \text{ with } \phi(g)1\}$. Let $g, k \in K$ then by the rules for homomorphisms if

    a) $\phi(gk) = \phi(g)\phi(k) = 1 \times 1 = 1$, so $gh \in K$;

    b) $1 = \phi(gg^{-1}) = \phi(g)\phi(g^{-1})$. So $\phi(g_{-1}) = (\phi(g))^{-1} \in K$;

    c) Let $g \in G$ and $k \in K$. Then $\phi(g^{-1}kg) = \phi(g^{-1}\phi(k)\phi(g) = (\phi(g))^{-1}\phi(g) = 1$. So K is normal.

(ii) By the first item above we see that $\phi(g^k) = \phi(g)^k$ for any $k \in \mathbb{Z}$. So if $g^n = 1$ then $\phi(g)^n = 1$. The result now follows.

4.4 This is a simple application of Lagrange's Theorem. Any subgroup has to have order dividing p, where p is the order of the group. So any subgroup has order 1 or p.

4.5 Let G and H have order p. Choose a an element $g \in G$ with $g \neq 1$ and $h \in H$ with $h \neq 1$. By the previous $G = \langle g \rangle$ and $H = \langle h \rangle$. Now map $g \to h$. This gives the required isomorphism.

4.6 Simple application of Lagrange as 5 does not divide 24.

4.7 Simple application of Lagrange as p does not divide $n!$ if $p < n$.

4.8

Cycle type	Order	Number
1^5	1	1
$2^1, 1^3$	2	10
$2^2, 1^1$	2	15
$3^1, 1^2$	3	20
$3^1, 2^1$	6	20
$4^1, 1^1$	4	30
5^1	5	24

4.9 This just involves looking at the possible cycle lengths. We need a partition of $10 = a+b+c+\cdots$ so that $\text{lcm}(a,b,c,\ldots)$ is maximal. It is $5+3+2$.

4.10 We see that the number of transpositions in the symmetric group of degree r is just $\frac{r(r-1)}{2}$ So the *ogf* is given by $\sum_{r\geq 2}\frac{r(r-1)}{2}z^r$ so the answer is $\frac{z^2}{(1-z)^3}$.

4.11 The main thing to check is that if g is a permutation there is a permutation h so that $gh = hg =$ the identity map. For any $x \in X$ define $h(x) = y$ where $g(y) = x$. Since g is onto there is such a y and because g is $1-1$ then y is unique.

4.12 Define the action of G on subsets of G by $g(X) = gX$ where $X \subseteq G, g \in G$. This is a well-defined action and the set $G_X = \{G : gX = X\}$ is just the stabilizer of a point in the action.

4.13 Consider the set $\{g^m : m \in \mathbb{Z}\}$. Note that the number of elements in this set n. So amongst the set $\{g^0,g^1,\ldots,g^n\}$ there must be two elements the same. Say $g^i = g^j$ for $0 \leq i,j \leq n$. Then $g^{j-i} = 1$. So there are positive natural numbers m so that $g^m = 1$. Let m be the smallest. We now claim that $\{g^0,g^1,\ldots,g^{m-1}\} = \langle g\rangle$ and hence that $m = n$ as claimed. Let $k \in \mathbb{Z}$. We can write $k = am+r, 0 \leq r < m$. Then

$$g^k = g^{am+r} = g^{am}g^r = (g^m)^a g^r = g^r.$$

So $g^k \in \{g^0,g^1,\ldots,g^{m-1}\}$.

Now let $g^d = 1$ and put $d = am+r, 0 \leq r < m$. As above $g^d = g^r = 1$ but by the previous paragraph we see that $r = 0$.

4.14 We first show that it has order 24. It is clear that the action is transitive on vertices. So fix any vertex, say V, then the stabilizer of V acts on the triangle formed by the remaining vertices. So the stabilizer is isomorphic to $\text{Sym}(3)$ and has order 6. There are four vertices so the stabilizer of V has index 4 in the whole group. So we have order 24. But clearly the whole group is inside $\text{Sym}(4)$ as there are four vertices.

An alternative approach is geometric. Consider the various rotations that will preserve the tetrahedron. For example choose any vertex and the midpoint of the opposite face, then rotate through $\pi/3$. By choosing each vertex in turn there are eight distinct elements of order 3. Then consider reflections.

4.15 The key argument is that if g and h are permutations and g has a cycle $(a_1a_2\ldots a_r)$ then hgh^{-1} has a cycle $(h(a_1)h(a_2)\ldots h(a_r))$.

If $(b_1b_2\ldots b_r)$ is another cycle, consider a permutation k that has $k(a_i) = b_i$. Then $k(a_1a_2\ldots a_r)k^{-1} = (b_1b_2\ldots b_r)$. So putting disjoint cycles together gives the result.

4.16 The subgroup that fixes $X_1 + X_2 - X_3 - X_4$ is $\{(12),(34),(12)(34)\}$. One such polynomial is $X_1X_2 + X_2X_3 + X_3X_4 + X_4X_1$.

4.17 An element of order p generates a cyclic group of order p. So each orbit has length 1 or p. Since n is not divisible by p there has to be an orbit of length 1. An orbit of length 1 corresponds to a fixed point.

4.18 Using the definitions $\phi(g)\phi(g^{-1}) = \phi(1)$. So $\phi(g^1) = (\phi(g))^{-1}$. Now it is straightforward to show that ϕ is a bijection.

4.19 Using the definitions $\phi(g)\phi(h) = \phi(g)(xh^{-1}) = xh^{-1}g^{-1} = x(gh)^{-1} = \phi(gh)(x)$ which is exactly what is required. The identity works correctly.

The reason for the inverse is to make the product come out correctly, without the inverse we would have $\phi(g)\phi(h) = \phi(hg)$.

4.20 Let $x \in \{1, 2, \ldots, n\}$. Then start with x and keep applying g. Any element in G is of the form g^m for some $m \in \mathbb{N}$. So y is in the same orbit as x if and only if $y = g^k(x)$ for some $k \in \mathbb{N}$.

4.21 We can use a similar argument to that used for the tetrahedron. There are six faces and the symmetry group, say G, acts transitively on them. Pick any face, say F, and note that $[G : G_F] = 6$. The opposite face has to be fixed. Now the remaining four faces can be thought of as a square and so $|G_F| = 8$. so the order is 48.

Geometrically the elements are: 9 rotations (by $\pm\pi/2$ or π) about centres of 3 pairs of opposite faces, 6 rotation (by π) about centres of 6 pairs of opposite edges. 8 rotations (by $\pm 2\pi/3$) about 4 pairs of opposite vertices (diagonals). [8] Together with the identity this accounts for all 24 elements. Then there are the reflections.

4.22 Consider the class

$$|G| = \sum_1^k |C_i|.$$

Divide everything through by $|G| = n$ to get the equation

$$1 = \frac{1}{n} + \frac{n_1}{n} + \cdots + \frac{n_r}{n}.$$

If there are only two conjugacy classes then we have the equation

$$1 = \frac{1}{n} + \frac{k}{n}.$$

So $k = n - 1$ but k divides n so $k = 1$ and $n = 2$.

If there are only three conjugacy classes then we have the equation

$$1 = \frac{1}{n} + \frac{k}{n} + \frac{l}{n}.$$

So $k + l = n - 1$ and both k and l divide n. We can list the possibilities:

(i) $k = l = 1$ and $n = 3$

(ii) $k = 3, l = 2$ and $n = 6$.

4.23 This is a simple application of the Orbit-Stabilizer Theorem. Since every orbit has order the index of a subgroups and such indices divide the order of the group (Lagrange's Theorem), the result follows.

4.24 For $r = 2$ there are just the two orbits: $\{(x,y) : x \neq y\}$ and $\{(x,x)\}$.

For $r = 3$ we have the orbits given by $\{(x_1,x_2,x_3) : x_i \text{ all distinct}\}$, $\{(x_1,x_1,x_2) : x_1 \neq x_2\}\{(x_1,x_2,x_1) : x_1 \neq x_2\}$, $\{(x_2,x_1,x_1) : x_1 \neq x_2\}$ and finally $\{(x_1,x_1,x_1)\}$.

The number of orbits is determined by the positions of the entries which are the same. So we can construct a partition of r where each subset of the partition is the positions where the entries are the same. So for $r = 3$ we have the partitions, $\{1\},\{2\},\{3\}$, $\{1,2\},\{3\}$, $\{1,3\},\{2\}$, $\{2,3\},\{2\}$ and $\{1,2,3\}$.

4.25 The first term of the sum counts how often p divides n or how many of the sequence $\{1,2,\ldots,n\}$ are divisible by p. Similarly the second term counts how often p^2 divides n or how many of the sequence $\{1,2,\ldots,n\}$ are divisible by p^2. Adding all the terms gives how many ps divides $n!$, which is the order $\text{Sym}(n)$.

4.26 We know $\binom{p^a m}{p^a} = \frac{(mp^a)!}{p^a!(mp^a - m)!}$. By the previous exercise $\binom{p^a m}{p^a}$ is divisible by

$$mp^{a-1} + mp^{a-2} + \cdots + m - (p^{a-1} + p^{a-2} + \cdots + 1)$$
$$- ((m-1)p^{a-1} + (m-1)p^{a-2} + \cdots + (m-1)) = 0.$$

Chapter 5

5.1 We start from the Binomial theorem slightly adapted, $\sum_{k \geq 0} \binom{r}{k} z^{k+1} = z(1+z)^r$. If we differentiate this (wrt z) and let $z = 1$ the result follows.

5.2 Think of choosing a football team of k players, with a captain, from r players. We can choose the k from r and then 1 from the k; or we can choose 1 from r and then $k-1$ from the $r-1$ remaining.

5.3 We may choose k from $r+1$ in $\binom{r+1}{k}$ ways. Now consider an arbitrary element. Either it is included in the choice, so choose the remainder $k-1$ from r in $\binom{r}{k-1}$ ways; or it is not included, so choose k from r remaining in $\binom{r}{k}$ ways, and hence the result. Three successive diagonal sums of Pascal's triangle have each term of this recurrence, so they also obey it. As the first two sums are Fibonacci, so are they all.

5.4 Think of a committee of k members and an executive sub-committee of l members – see Exercise 5.2.

5.5 Split up r into m and $r-m$. We can choose 0 from m and k from $r-m$; or 1 from m and $k-1$ from $r-m$; etc.

5.6 We have $rc_{r-1} = r\sum_{k \geq 0} \binom{r-1}{k} k! = \sum_{k \geq 0} \binom{r}{k+1}(k+1)! = \sum_{k \geq 0} \binom{r}{k} k! - 1 = c_r - 1$ using Exercise 5.2. In constructing a c_r from a c_{r-1} the "new" button r either joins an existing combination (and it has r places to do this) or it is a combination by itself.

5.7 From the last exercise, we have $c_r = rc_{r-1} + 1$ and $c_{r+1} = (r+1)c_r + 1$. Subtracting these gives the result.

5.8 The result is true for $r = 1$. Suppose it is true for r. Then $r! = \sum_{k \geqslant 0}(-1)^{r-k}\binom{r}{k}c_k$ and hence $(r+1)! = \sum_{k \geqslant 0}(-1)^{r-k}(r+1)\binom{r}{k}c_k$. So

$$(r+1)! = \sum_{k \geqslant 0}(-1)^{r-k}\binom{r+1}{k+1}(k+1)c_k = \sum_{k \geqslant 0}(-1)^{r-k}\binom{r+1}{k+1}(c_{k+1}-1)$$

$$= \sum_{k \geqslant 0}(-1)^{r+1-k}\binom{r+1}{k}c_k$$

and the induction is complete.

5.9 The answers appear in the following section.

5.10 In general, $\sum_{l \geqslant 0} 2^{l-r}\binom{r}{l}2^{k-l}\binom{l}{k} = 2^{k-r}\sum_{l \geqslant 0}\binom{r}{l}\binom{l}{k} = 2^{k-r}2^{r-k}\binom{r}{k} = \binom{r}{k}$, so the result is true.

5.11 The characteristic equation of the matrix P_3 is $(1-\lambda)^3 = 0$. Every matrix satisfies its own characteristic equation and the result follows. In general $(P_n - I)^n = 0$.

5.12 (i) $\{B_r\} = \{1, 1, 2, 5, 15, \ldots\}$. (ii) Use Theorem 5.12.

5.13 We have $\sum_{r \geqslant 0}(-1)^r r^2 z^r = \frac{-z+z^2}{(1+z)^3}$. If we multiply each side by $\frac{1}{1-z}$ then on the left we obtain the desired sum, and on the right we have $\frac{-z+z^2}{(1-z)(1+z)^3}$. Partial fractions on this gives $\frac{1}{(1+z)^3} - \frac{1}{(1+z)^2}$. It now follows that $\sum_{k=0}^{n}(-1)^k k^2 = (-1)^n\binom{n+2}{2} - (-1)^n\binom{n+1}{1} = (-1)^n\frac{n(n+1)}{2}$.

5.14 We have $L_{r+2} = \phi_1^{r+2} + \phi_2^{r+2}$ and so $\sum_{r \geqslant 0}L_{r+2}z^r = \frac{\phi_1^2}{1-\phi_1 z} + \frac{\phi_2^2}{1-\phi_2 z} = \frac{L_2+z}{1-z-z^2} = \frac{3+z}{1-z-z^2}$. (i) We have $\frac{1}{1-z}\sum_{r \geqslant 0}L_r z^r = \frac{2-z}{(1-z)(1-z-z^2)} = -\frac{1}{1-z} + \frac{3+z}{1-z-z^2}$ and hence the result.(ii) So $\frac{1}{3+z}\sum_{r \geqslant 0}L_{r+2}z^r = \frac{1}{1-z-z^2} = \frac{1}{2}\sum_{r \geqslant 0}F_r z^r$ and hence the result.

5.15 See previous question. We have $\frac{1}{1-2z}\sum_{r \geqslant 0}L_r z^r = \frac{2-z}{(1-2z)(1-z-z^2)} = \frac{6}{1-2z} - \frac{4+3z}{1-z-z^2}$ and hence the result.

5.16 We have $\frac{z}{(1-z^2)(1-z-z^2)} = \frac{1}{1-z^2}\sum_{r \geqslant 0}F_r z^r = \frac{1}{(1-z-z^2)} - \frac{1}{2}\left(\frac{1}{1+z} + \frac{1}{1-z}\right)$ and hence the results. The corresponding Lucas results are $\sum_{k=0}^{r}L_{2k+1} = L_{2r+2} - 2$ and $\sum_{k=0}^{r}L_{2k} = L_{2r+1} + 1$.

5.17 We have $\left(\sum_{r \geqslant 0}F_r z^r\right)^2 = \frac{1}{5}\left(\frac{1}{1-\phi_1 z} - \frac{1}{1-\phi_2 z}\right)^2 = \frac{1}{5}\left(\frac{1}{(1-\phi_1 z)^2} - \frac{2}{1-z-z^2} + \frac{1}{(1-\phi_2 z)^2}\right)$ and hence $\sum_{k=0}^{r}F_{r-k}F_k = \frac{1}{5}((r+1)L_r - 2F_{r+1})$. Similarly, $\sum_{k=0}^{r}L_{r-k}L_k = (r+1)L_r + 2F_{r+1}$.

5.18 If we use the zD operator on the generating function of the Fibonacci sequence we obtain $\sum_{r \geqslant 0}rF_r z^r = \frac{z+z^3}{(1-z-z^2)^2}$; the denominator of this provides the recurrence $u_r = 2u_{r-1} + u_{r-2} - 2u_{r-3} - 2u_{r-4}$.

5.19 The result leads to $\frac{1}{1+z}\sum_{r\geqslant 0}F_{r+2}z^r = \frac{1}{1-z-z^2}$ and hence $\sum_{k=0}^{r}(-1)^{r-k}F_{k+2} = F_{r+1}$.

5.20 We have $\sum_{r\geqslant 0}(L_r - F_r)z^r = \frac{2(1-z)}{1-z-z^2}$ and dividing each side by $1-z$ gives the result. We also have $5F_r = \sqrt{5}(\phi_1^r - \phi_2^r) = (\phi_1 - \phi_2)(\phi_1^r - \phi_2^r)$ and hence the result.

5.21 We start with Theorem 2.3: $\sum_{k=0}^{n}a^r = \frac{a^{n+1}-1}{a-1}$. If we apply the Euler operator aD_a to this we obtain $\sum_{k=0}^{n}ra^r = \frac{(a-1)(n+1)a^n - (a^{n+1}-1)}{(a-1)^2}$ and hence the result.

5.22 (i) We have $k^2 = k(k-1) + k = 2\binom{k}{2} + \binom{k}{1}$. So

$$\sum_{k=0}^{n}k^2 = 2\sum_{k=0}^{n}\binom{k}{2} + \sum_{k=0}^{n}\binom{k}{1}$$

and by Theorem 5.3 this is

$$\sum_{k=0}^{n}k^2 = 2\binom{n+1}{3} + \binom{n+1}{2} = \frac{n(n+1)(2n+1)}{6}.$$

(ii) We have $\sum_{r\geqslant 0}r^2 z^r = \frac{z+z^2}{(1-z)^3}$ and so multiplying each side by $1/(1-z)$

$$\sum_{k=0}^{n}k^2 = \binom{n+2}{3} + \binom{n+1}{3} = \frac{n(n+1)(2n+1)}{6}.$$

5.23 In choosing r from $2r$ we can choose k from r and $r-k$ from r, where $0 \leq k \leq r$. So

$$\binom{2r}{r} = \sum_{k\geqslant 0}\binom{r}{k}\binom{r}{r-k} = \sum_{k\geqslant 0}\binom{r}{k}^2.$$

5.24 The diagonal sums operate like those in Pascal's triangle (Exercise 5.3). The first sum is 2 and the second 1, and hence the diagonal sums are Lucas numbers.

Chapter 6

6.1 By Theorem 6.4 we have

$$1 = \frac{1}{|G|}\sum_{g\in G}|\mathrm{Fix}(g)|.$$

But $|\mathrm{Fix}(1)| = n$ where the degree of G is n. So if $|\mathrm{Fix}(g)| \geq 1$ for all $g \in G$ we get $|G| \leq n + |G| - 1$ which can only happen if $n = 1$.

6.2 Using the same equation this time we find $k|G| = n = |X|$. So $|G|$ divides $|X|$. Note that if $k = 1$ then $|X| = |G|$.

6.3 The two sets of vertices $\{A,B,D,E\}$ and $\{F,C\}$ are both fixed. We can write the elements as $\{(A)(B)(C)(D)(E)(F),(AB)(ED)(FC),(AE)(BD),(AD)(BE)(FC)\}$. There are two orbits on vertices as above; check using Equation (6.3), $\frac{6+2+0+0}{4} = 2$.

There are eight edges and this time there are three orbits $\frac{8+2+2+0}{4} = 3$, and they have sizes $2, 2$ and 4.

6.4 There are two orbits, each of size 4.

6.5 We again use the orbits and fixed point theorem. The only sequence that can be fixed is one all of whose entries are the same. So $k = \frac{n^p + (p-1)n}{p}$, where k is the number of orbits. So $p|(n^p + (p-1)n)$, which gives the result.

6.6 Using the results from the solution 4.21, where we only consider the rotations, so the group has order 24. The conclusion is that the cycle index of the group is

$$\frac{1}{24}\left(Z_1^6 + 6Z_1^2Z_4 + 3Z_1^2Z_2^2 + 8Z_3^2 + 6Z_2^3\right).$$

Now replace Z_i with 2 and get the answer 14.

6.7 The cycle index polynomial for the alternating group of degree four is $\frac{1}{12}(Z_1^4 + 3Z_2^2 + 8Z_3Z_1)$.

6.8 $\frac{1}{4}(Z_1^6 + 2Z_2^3 + Z_2^2Z_1^2)$. So with k colours there are $\frac{1}{4}(k^6 + 2k^3 + k^4) = \frac{1}{4}(k^3(k+1)(k^2 - k+2))$.

6.9 For vertices and faces the answer will be the same. There are four of each and the group acts like the symmetric group of degree 4. Using Table 4.1,

$$P_G(Z_1,Z_2,Z_3,Z_4) = \frac{1}{24}\left(Z_1^4 + 3Z_2^2 + 8Z_3Z_1 + 6Z_2Z_1^2 + 6Z_4\right).$$

With two colours there are five ways and with three colours there are 15 ways.

To consider the action of edges, recall that there are six edges so

$$P_G(Z_1,Z_2,Z_3,Z_4,Z_5,Z_6) = \frac{1}{24}\left(Z_1^6 + 3Z_2^2Z_1^2 + 8Z_3^2 + 6Z_2^2Z_1^2 + 6Z_4Z_2\right).$$

With two colours there are 11 ways and with three colours there are 66 ways.

6.10

$$P_G(Z_1,Z_2,Z_3,Z_4,Z_5,Z_6) = \frac{1}{8}\left(Z_1^6 + 5Z_2^2Z_1^2\right).$$

6.11 The required cyclic polynomial is:

$$P_G(Z_1,Z_2,Z_3) = \frac{1}{6}\left(Z_1^3 + 2Z_3 + 3Z_1Z_2\right).$$

Now replace Z_i with $(R^i + G^i)$ to get

$$\frac{1}{6}\left((R+G)^3 + 2(R^3 + G^3) + 3(R+G)(R^2 + G^2)\right).$$

The total number is given by putting $R + G = 1$ (note that this is the same as substituting 2 for each Z_i).

To find the number with two reds is the coefficient of R^2G in the above polynomial. There are only two terms which can contribute, $(R+G)^3$ and $3(R+G)(R^2+G^2)$, and this gives $\frac{6}{6} = 1$.

6.12 We have already worked out the cycle structure (see Table 6.1). So

$$P_G(Z_1, Z_2, Z_3, Z_4) = \frac{1}{8}\left(Z_1^4 + 3Z_2^2 + 2Z_2Z_1^2 + 2Z_4\right).$$

Let the colours be R, G and B. Then the polynomial we need to study is

$$\frac{1}{8}\left((R+G+B)^4 + 3(R^2+G^2+B^2)^2 + 2(R^2+G^2+B^2)(R+G+B)^2 + 2(R^4+G^4+B^4)\right).$$

We are only interested in the terms which involve all the variables R, G and B. After some simplification the answer emerges as $\frac{24}{8} = 3$.

6.13

Element	Cycle structure on unordered pairs	Number
(1)(2)(3)(4)	1^6	1
(1234)	2^14^1	6
(123)	3^2	8
(12)(34)	2^21^2	3
(24)	2^21^2	6

So

$$P_G(Z_1, Z_2, Z_3, Z_4) = \frac{1}{24}\left(Z_1^4 + 3Z_2^2 + 8Z_3Z_1 + 6Z_2Z_1^2 + 6Z_4\right).$$

The same as the answer for the tetrahedron on edges!

The second part just doubles everything.

6.14 To do this observe that the group is of order 10. It has a rotation of order 5 and five reflections.

6.15 Assume that G has at least two orbits. Let Z be an orbit of G such that $|Z| < |X|$. Let $z \in Z$ and note that since P_z fixes exactly one point $|Z| \equiv 1 \pmod{p}$. Now choose $x \in X \setminus Z$. Then P_x fixes no element in Z so $|Z| \equiv 0 \pmod{p}$. This is a contradiction.

6.16 The non-trivial elements of the cyclic group act on itself as a p-cycle. For example $g \cdot g^i = g^{i+1}$. So the group has the identity element and $p - 1$ elements which consist of a p-cycle. So

$$P_G = \frac{1}{p}\left(Z_1^p + (p-1)Z_p\right).$$

Now colour the group with k colours. We can do this in $\frac{1}{p}(k^p + (p-1)k)$ ways. This has to be an integer.

6.17 We know from an earlier exercise that the required polynomial is:

$$\frac{1}{24}\left(Z_1^6 + 6Z_1^2 Z_4 + 3Z_1^2 Z_2^2 + 8Z_3^2 + 6Z_2^3\right).$$

Now replace Z_i with $R^i + W^i + B^i$ and then find the coefficient of $R^2 W^2 B^2$. This is 6.

Chapter 7

7.1 We require the coefficient of $z^4/4!$ in the *egf* $(1+z)^6(1+z+z^2/2)$. This is $1,020$.

7.2 (i) Use Binet's formula: $\sum_{r\geq 0} \frac{L_r z^r}{r!} = \sum_{r\geq 0} \frac{\phi_1^r z^r}{r!} + \sum_{r\geq 0} \frac{\phi_2^r z^r}{r!} = e^{\phi_1 z} + e^{\phi_2 z}$. (ii) $F_{2r} = \frac{1}{\sqrt{5}}\left(\phi_1^{2r} - \phi_2^{2r}\right)$ and so $\sum_{r\geq 0} \frac{F_{2r} z^r}{r!} = \frac{1}{\sqrt{5}}\left(e^{\phi_1^2 z} - e^{\phi_2^2 z}\right)$.

7.3 We find $A = 1$, $B = 3$ and $C = 1$. So the *egf* is $(z + 3z^2 + z^3)e^z$.

7.4 The first form seems to lead nowhere; the second form leads to $T'(z) - t_1 = T(z) - t_0 + zT(z)$ and hence $T'(z) = (1+z)T(z)$. This differential equation has the solution $T(z) = e^{z + \frac{z^2}{2}}$.

7.5 Suppose the required *egf* is $C(z)$; using the three-step recipe gives the differential equation (see the second part of Example 7.8) $C(z) - c_0 = zC(z) + e^z - 1$ and so the *egf* is $C(z) = \frac{e^z}{1-z}$.

7.6 The two *egfs* are $\frac{1}{1-z}$ and e^{-z}. They represent the sequences $\{r!\}$ and $\{(-1)^r\}$ respectively. So by the product formula, we have $d_r = \sum_{k\geq 0}\binom{r}{k}(r-k)!(-1)^k$.

7.7 We may rearrange the *egf* as $e^{-z}\sum_{r\geq 0}\frac{t_r z^r}{r!} = e^{\frac{z^2}{2}}$ and if we expand each side the coefficients of odd powers of z on the right are zero. Hence so are those on the left. Using Theorem 7.12 gives the required result. For even powers of z, we have $\sum_{k\geq 0}(-1)^k\binom{2r}{k}t_k = \frac{1}{2^{2r}}$.

7.8 This comes from the square of the *egf* of the Lucas numbers

$$(L(z))^2 = \left(e^{\phi_1 z} + e^{\phi_2 z}\right)^2 = e^{2\phi_1 z} + 2e^z + e^{2\phi_2 z} = L(2z) + 2e^z.$$

Then comparing coefficients of z^r on each side and making use of the product formula gives the result.

7.9 The sequences whose *egfs* make up the product are $\{1\}$ and $\{r!\}$ respectively. By the product formula we now have the *egf* of the push-button lock sequence as the function $\frac{e^z}{1-z}$.

7.10 We write the *egf* as the product of two functions $1/(1-z)^2$ and e^z. The first of these is the *egf* of the sequence $\{(r+1)!\}$ and the second is the *egf* of the sequence $\{1\}$. The product formula now gives the required result.

7.11 (i) Differentiating the *egf* gives $\sum_{r\geq0}\frac{\sigma_{r+1}z^r}{r!} = ze^{\frac{z^2}{2}} = z\sum_{r\geq0}\frac{\sigma_r z^r}{r!}$ and hence the result. (ii) Now $\sigma_{0,=2} = 1$ and so $\sigma_{2,=2} = 1.1$, $\sigma_{4,=2} = 1.1.3$ and the result follows.

7.12 (i) From Theorem 4.17 we have $\sigma_{3r,=3} = \frac{(3r)!}{3^r r!}$ (and zero otherwise), so

$$\sum_{r\geq0}\frac{\sigma_{r,=3}z^r}{r!} = \sum_{r\geq0}\frac{(3r)!z^{3r}}{3^r r!(3r)!} = \sum_{r\geq0}\frac{z^{3r}}{3^r r!} = e^{\frac{z^3}{3}}.$$

(ii) We can choose k from r and make them into three cycles in $\binom{r}{k}\sigma_{k=3}$ ways, and then make the remaining terms into involutions. Hence, using Theorem 7.12, we have $\sum_{r\geq0}\frac{\sigma_{r,\leq3}z^r}{r!} = \left(\sum_{r\geq0}\frac{\sigma_{r,=3}z^r}{r!}\right)\left(\sum_{r\geq0}\frac{\sigma_{r,\leq2}z^r}{r!}\right)$ and the result follows. In the final part, the given function can be thought about in two ways. First it is the *egf* for the number of permutations whose cycles are of length 1 or 2 or 3, and so on. In other words *any* permutation. On the other hand it is also $e^{-\log(1-z)} = \frac{1}{1-z}$ which is the *egf* for the sequence which also counts all permutations. Reassuringly, these are the same.

7.13 Differentiating the *egf* twice gives $\sum_{r\geq0}\frac{\sigma_{r+2,=2}z^r}{r!} = (1+z^2)e^{\frac{z^2}{2}}$ and hence the result.

7.14 We have $c_r = \sum_{k\geq0}\binom{r}{k}k! = r!\sum_{k=0}^r\frac{1}{k!} = r!\left(\sum_{k\geq0}\frac{1}{k!} - \sum_{k\geq r+1}\frac{1}{k!}\right) = r!e - Q_r$ where $|Q_r| < \frac{1}{r}$. We deduce that $\frac{c_r}{r!} = e - \frac{Q_r}{r!} \to e$ as $r \to \infty$.

7.15 We have $r! = \sum_{k\geq0}\binom{r}{k}\sigma_{k,\neq2}\sigma_{r-k,=2}$ and then using Theorem 7.12 this leads to the result given. If we differentiate this we obtain $\sum_{r\geq0}\frac{\sigma_{r+1,\neq2}z^r}{r!} = \frac{-ze^{-\frac{z^2}{2}}}{1-z} + \frac{e^{-\frac{z^2}{2}}}{(1-z)^2} = -z\sum_{r\geq0}\frac{\sigma_{r,\neq2}z^r}{r!} + \frac{1}{1-z}\sum_{r\geq0}\frac{\sigma_{r,\neq2}z^r}{r!}$ which leads to the given recurrence.

7.16 The *egf* for the sequence $\{c_r\}$ is $\frac{e^z}{1-z} = e^{2z}\left(\frac{e^{-z}}{1-z}\right) = e^{2z}\sum_{r\geq0}\frac{d_r z^r}{r!}$ and the result follows by Theorem 7.12.

7.17 (i) We have $o_{r+1} = \sum_{k\geq0}\binom{r}{k}o_k o_{r-k}$ and so $o_2 = \sum_{k\geq0}\binom{1}{k}o_k o_{1-k} = o_0 o_1 + o_1 o_0 = 0$; $o_3 = o_0 o_2 + 2o_1 o_1 + o_2 o_0 = 2$; $o_4 = 0$; $o_5 = 16$; etc.
(ii) Similarly, $e_2 = \sum_{k\geq0}\binom{1}{k}o_k e_{1-k} = o_0 e_1 + o_1 e_0 = 1$; $e_3 = o_0 e_2 + 2o_1 e_1 + o_2 e_0 = 0$; $o_4 = 5$; $o_5 = 0$; etc.

7.18 (i) The (sort of) *egf* may be written $(1-z)\sum_{r\geq0}\frac{u_r z^r}{r!} = \frac{e^{-z}}{1-z} = \sum_{r\geq0}\frac{d_r z^r}{r!}$ and the result follows. (ii) If we differentiate the given *egf* then $\sum_{r\geq0}\frac{u_{r+2}z^r}{r!} = -\frac{e^{-z}}{(1-z)^2} + 2\frac{e^{-z}}{(1-z)^3} = -\sum_{r\geq0}\frac{u_{r+1}z^r}{r!} + \frac{2}{1-z}\sum_{r\geq0}\frac{u_r z^r}{r!}$ and the recurrence follows. The sequence begins $\{1,1,3,11,53,\cdots\}$.

7.19 Differentiating the *egf* gives

$$\sum_{r\geq 0} \frac{\sigma_{r+1,>2}z^r}{r!} = -(1+z)\sum_{r\geq 0} \frac{\sigma_{r,>2}z^r}{r!} + \frac{1}{1-z}\sum_{r\geq 0} \frac{\sigma_{r,>2}z^r}{r!}$$

and the recurrence follows.

7.20 Again, we recognize the left-hand side of this as a derivative, and the right as the product of two *egf*s. It follows that $E'(z) = O(z)E(z)$. This is easily integrated and gives $E(z) = \sec z$ as required.

Chapter 8

8.1 Since the sum of the degrees of the vertices is even if each vertex has odd degree then there has to be an even number of them.

8.2 We begin by observing that $\sum d(v) = 2e$. If $d(v) \geq 2$ for all v then $2e = \sum d(v) \geq 2|V|$. So $e > |V|$, which contradicts $|V| > e$.

8.3 If a vertex has $d(v) = 0$ then it is not connected to anything.

8.4 Let $(v_0, e_1, v_2, \ldots, e_n, v_n)$ be a path. We will use induction on n to get the result. Note that if $n = 1$ there is nothing to prove.

Assume that we have a path as above and we assume that $v_i = v_j$. If $i \neq 0$ and $j \neq n$ then we can just omit the section from v_i to v_j and have a shorter path. Then induction gives the result.

8.5

(i) This is clear since each vertex in U can be joined to at most $|W|$ vertices. So we get $m \leq rs$.

(ii) $n = r + s$ so we want to maximize the expression rs. This happens when $r = s = n/2$.

8.6 This is clear. If there were more than two components choose u, v, w in three distinct components. But an edge can only join two vertices, so if $e = (u, v)$ then the component containing w is still a connected component.

8.7 If there are two components at leat one of them has $\leq r/2$ vertices. A vertex in such a component can be connected to at most $r/2 - 1$ vertices.

8.8 Let u and v be the pair whose degree sum is greater than $r - 1$. By counting it can be seen that any other vertex is joined to either u or v.

8.9 We must first write Menger's Theorem in the edge form:
Let $\Gamma = (V, E)$ be a graph and let v and w be two connected vertices of V. Let k be the

size of the smallest set F of edges with the property that v and w are not connected in $\Gamma \setminus F$. Let l be largest number of simple paths from v to w which only have v and w in common. Then $k = l$.

If we start with the set of edges F in the line graph it becomes a set of vertices which disconnects in the same way.

8.10 Choose and vertex of degree 1, say v. Then set the partitions according as the distance from v is odd or even. Or simply observe that a tree has no cycles of odd length.

8.11 Let T be a tree with n vertices. Then it has $n - 1$ edges, so if it has exactly two vertices of degree 1 the rest have degree 2, as $\sum d(v) = 2n - 2$.

Let T_1 and T_2 be two such graphs. Choose vertices of degree 1, say u_1 and u_2, of T_1 and T_2 respectively. These are joined to a unique vertex in their respective graphs. Then since these have degree 2 there is only one other vertex they are connected to. So proceeding in this way we construct the required isomorphism.

Note that these trees have an automorphism group of order 2.

8.12 A cycle is just a graph which is one simple closed path. To show that a graph with all vertices of degree 2 is a simple closed path start anywhere, say v_1. Choose v_2 adjacent to v_1 – there are only two choices. Then choose v_3 to be the remaining vertex adjacent to v_2. Keep going till you get to v_n and then the only edge not used would be the edge (v_n, v_1), and there is the simple closed path.

Such a path has an automorphism group of order $2n$, so there are $\frac{n!}{2n}$ ways to label it. So the *egf* is

$$\sum \frac{n!}{2n} \times \frac{1}{n!} z^n = \frac{1}{2} \log(1 - z).$$

8.13 One, and two.

8.14 Two.

8.15 We look at the table of elements of $\mathrm{Sym}(5)$ acting on pairs.

Cycle type	Action on pairs	Number
1^5	1^{10}	1
$2^1, 1^3$	$1^4 2^3$	10
$2^2, 1^1$	$1^2 2^4$	15
$3^1, 1^2$	$1^1 3^3$	20
$3^1, 2^1$	$1^1 3^1 6^1$	20
$4^1, 1^1$	$2^1 4^2$	30
5^1	5^2	24

Now calculate the cycle index and substitute 2 for all indeterminates. This gives the answer 34.

8.16 We use the equation $v - e + f = 2$. Every face is bounded by a cycle so if the shortest cycle has length g and every edge is the boundary of two faces, be careful about counting. So $fg \geq 2e$. Substituting into the equation above gives the right answer.

8.17 Since the sum of the degrees of the vertices is twice the number of edges, if every vertex has degree at least 6 we get $6v \leq 2e$. Also $3f \leq 2e$, since every face has at least three edges in the boundary. So $12 + 6e = 6v + 6f \leq 6e$, which is clearly a contradiction.

8.18 This is again a variation on a theme. We get $3v \leq 2e$ and $5f \leq 2e$. So $30 + 15e = 15v + 15f \leq 16e$. So $e > 30$ and $f\frac{60}{5} = 12$.

8.19 This time let Γ be the graph and Γ' be its complement. The have the same number of vertices, say v and if one has e edges the other has $\binom{v}{2} - e$ edges. Let them have f and f' faces respectively. So we get the equations $v - e + f = 2$ and $v - (\binom{v}{2} - e) + f' = 2$. Combining this with the observation that $3f \leq 2e$ and $3f' \leq 2(\binom{v}{2} - e)$ gives the result, after some manipulation.

8.20 Consider the any face, say F, and assume that it has more than three vertices on its boundary. Then there are two which are not adjacent. Just join these two vertices – remember that edges do not have to be straight. The new graph will still be planar. So each face has three vertices and three edges on its boundary. So $3f = 2e$. Now use Euler's formula and get $6 = 3v - 3e + 3f = 3v - 3e + 2e = 3v - e$. The two results follow.

8.21 To say that the complement of a graph has no triangles is the same as saying that the graph does not have three vertices which form a null graph. Given any vertex v, consider the two subsets U and W which consist of the vertices adjacent to v and those not adjacent to v. One of these has to have size at least 3. If it is U and there are two adjacent vertices in U then there is a triangle. If there are not two adjacent vertices then U is a null graph.

We argue similarly if W has size at least three. If W contains two vertices which are not adjacent then together with v they form a null graph. Otherwise W has a triangle.

8.22 Choose any vertex v and count the number of vertices at each distance i from v. When $i = 1$ this is clearly bounded by m. For $i = 2$ the answer is $m(m - 1)$ and for arbitrary i the answer is $m(m - 1)^{i-1}$. So the bound on the number of vertices is

$$1 + m\left(\sum_1^d m(m - 1)^{i-1}\right) = 1 + m\frac{(m - 1)^d - 1}{m - 2}.$$

8.23 This is a straightforward application of Hall's Marriage Theorem 8.23.

8.24 Think of the pizza as a circle and consider the result of cutting the pizza into pieces as creating a planar graph. We do this to maximize the number of pieces so that there are no coincidences amongst the intersections. We need to count the number of vertices and edges. We must keep in mind the edge of the pizza as an edge of the graph which keeps being subdivided. Assume there are n cuts.

There are $2n$ vertices on the edge of the pizza and $\binom{n}{2}$ vertices inside the pizza.

There are $2n$ edges around the edge of the pizza, and as each cut will cut each other cut there are $n \times n$ internal edges.

Remembering that we are not interested in the infinite face we get the number of pieces as $\frac{2+4n+2n^2-n^2+n-4n}{2}$, using Euler's formula. So the answer is $\frac{n^2+n+2}{2}$. (Note that this question is also considered in Example 2.18.)

8.25 The cyclic index polynomial is

$$\frac{1}{48}\left((B+W)^8 + 13(B^2+W^2)^4 + 6(B+W)^4(B^2+W^2)^2 + \right.$$
$$\left. 8(B+W)^2(B^3+W^3)^2 + 12(B^4+W^4)^2 + 8(B^2+W^2)(B^6+W^6)\right).$$

This gives 22 different graphs.

Chapter 9

9.1 Any positive integer can be written as a denary string (decimal notation !). So $P(z) = 1/1 - z$ as the representation is unique.

9.2 The generating function in which each part appears at most 4 times is $\prod_{r\geq 1}\sum_{m=0}^{4} z^{rm}$, which simplifies to $\prod_{r\geq 1}\left(\frac{1-z^{5r}}{1-z^r}\right) = \prod_{r\not\equiv 0 \ (\text{mod } 5)}\frac{1}{1-z^r}$, which is the generating function for the partitions whose parts are not divisible by 5.

9.3 (i) $\prod_{k=1}^{5}\frac{1}{1-z^k}$; (ii) $\prod_{k>4}\frac{1}{1-z^k}$; (iii) $\prod_{k\geq 1}\frac{1}{1-z^{5k}}$.

9.4 The generating function is

$$\prod_{r\geq 1}\left(\frac{1}{1-z^r}-z^r\right) = \prod_{r\geq 1}\frac{1-z^r+z^{2r}}{1-z^r} = \prod_{r\geq 1}\frac{1+z^{3r}}{(1-z^r)(1+z^r)}$$

$$= \frac{1}{(1-z^2)(1-z^3)(1-z^4)(1-z^6)\cdots}$$

which is the generating function required.

9.5 The generating function is $\sum_{r\geq 0} p(r : \text{parts} \leq k)z^r = \frac{1}{(1-z)(1-z^2)\cdots(1-z^k)}$ and so

$$(1-z^k)\sum_{r\geq 0} p(r : \text{parts} \leq k))z^r = \frac{1}{(1-z)(1-z^2)\cdots(1-z^{k-1})} = \sum_{r\geq 0} p(r : \text{parts} \leq k-1)z^r$$

and the result follows. The enumeration starts with the observation that any partition *either* contains k so we can remove it and there are $p(r-k : \text{parts} \leq k)$ partitions left, *or* it doesn't contain k and there are $p(r :\leq k-1)$ such partitions. The result follows.

9.6 We have
$\sum_{r\geq 0} q_r z^r = \prod_{r\geq 0} \sum_{m=0}^{8} z^{3^r m} = \prod_{r\geq 0} \frac{1-z^{3^{r+2}}}{1-z^{3^r}} = \frac{1}{(1-z)(1-z^3)}$ and hence the required recurrence relation.

9.7 Any partition of r in which there are distinct parts *either* includes 1 *or* it does not. In the former case there are just $p(r-k : k-1$ distinct parts) [subtract 1 from each part and there are $k-1$ non-zero parts left]. In the latter case, there are just $p(r-k : k$ distinct parts) partitions [subtract 1 from each part and there are still k non-zero parts]. Hence the required recurrence.

9.8 Let $u_r = p(r : \text{even number of odd parts})$ then

$$\sum_{r\geq 0} u_r z^r = (1+z^{1+1}+\cdots)(1+z^{3+3}+\cdots)\cdots$$

$$= \frac{1}{(1-z^2)(1-z^6)(1-z^{10})}\cdots = \frac{1}{(1-z)(1+z)(1-z^3)(1+z^3)(1-z^5)(1+z^5)}\cdots$$

and hence

$$(1+z)(1+z^3)(1+z^5)\sum_{r\geq 0} u_r z^r = \frac{1}{(1-z)(1-z^3)(1-z^5)}\cdots$$

and hence the required recurrence, using Theorem 5.17, The generating function product.

9.9 We have

$$\sum_{r\geq 0} p(r : \text{distinct parts})z^r = (1+z)(1+z^2)(1+z^3)\cdots$$

$$= \frac{(1-z^2)(1+z^2)(1+z^3)\cdots}{(1-z)} = \frac{(1-z^4)(1+z^3)\cdots}{(1-z)}$$

$$= \frac{1}{1-z}\sum_{r\geq 0} p(r : \text{distinct parts, not a power of 2})z^r$$

and hence the required recurrence.

9.10 We have $\sum_{r\geq 0} p(r : \text{parts are powers of 2})z^r = (1+z)(1+z^2)(1+z^4)\cdots$ since each such power can be used either once or not at all (if it is used twice, that is the next

power). Now $(1+z)(1+z^2)(1+z^4)\cdots = \frac{(1-z^2)(1+z^2)(1+z^4)}{1-z}\cdots = \frac{(1-z^4)(1+z^4)}{1-z}\cdots = \frac{1}{1-z}$.
Hence, $p(r : \text{parts are powers of } 2) = 1$. (Obvious – r written in binary notation.)

9.11 $T_3 = \sum\limits_{k=0}^{2} T_k T_{4-k} = T_0 T_4 + T_1 T_3 + T_2 T_2 = 1; T_4 = \sum\limits_{k=0}^{3} T_k T_{5-k} = 2;$

$T_5 = \sum\limits_{k=0}^{4} T_k T_{6-k} = 5.$

9.12 Expanding $\sqrt{1-4z}$ by the Binomial Theorem gives $1 - 2z - 2z^2 - 4z^3 - 10z^4 \cdots$
and so the given generating function starts $z - z^2 - z^3 - 2z^4 \cdots$ which has negative
values of the triangular partition numbers as coefficients.

9.13 (i) $\sqrt{1-4z} = 1 - \sum\limits_{r \geq 1} \frac{1.1.3 \cdots (2r-3)2^{r-1}}{r!} z^r$ and so $T_r = \frac{1.1.3 \cdots (2r-5)2^{r-2}}{(r-1)!}$. (ii) We can
write this as $T_r = \frac{2^{r-2} \cdot (r-2)! 1.1.3 \cdots (2r-5)}{(r-2)!(r-1)!} = \frac{(2r-4)!}{(r-1)(r-2)!(r-2)!} = \frac{1}{r-1}\binom{2r-4}{r-2}.$

9.14 There will be a final multiplication of two brackets – one containing $1, 2, \ldots, k$, etc
terms and the other containing $r - k + 1$ terms. Hence $C_r = \sum_{k=0}^{r-1} C_k C_{r-k+1}.$

9.15 It is easy to prove that $C_r = \frac{1}{r+1}\binom{2r}{r} = \binom{2r}{r} - \binom{2r}{r+1}$ and as the latter two terms are
integers, so the given expression produces integers.

9.16 If we integrate this, we find that $\sum\limits_{r \geq 0} \frac{1}{r+1}\binom{2r}{r} z^{r+1} = \text{const} - \frac{\sqrt{1-4z}}{2}$. However, when
$z = 0$ we find that the value of the constant is $\frac{1}{2}$, which then leads to the required result.

9.17 When $z = \frac{1}{4}$ (for which value the power series expansion of the generating function is convergent) we have $\sum_{r \geq 0} \frac{C_r}{4^r} = 2.$

9.18 This result is simply the statement that any positive integer can be written
(uniquely) in base 3, that is as a unique partition whose parts are powers of 3. Algebraically, the first bracket may be written as $\frac{1-z^3}{1-z}$. When each bracket is written in
the corresponding way, there is a cascade of cancellation, leaving just $\frac{1}{1-z}$.

9.19 We have (from Exercise 9.6)

$$\sum_{r \geq 0} q(r)z^r = \frac{1}{(1-z)(1-z^3)} = \frac{1}{3}\left(\frac{1}{1-z} + \frac{1}{(1-z)^2} + \frac{1-z^2}{1-z^3}\right)$$

and hence $q(r) = \lfloor \frac{r}{3} \rfloor + 1.$

9.20 The number of specified partitions is

$$\sum_{r \geq 0} q(r)z^r = (1+z+z^2+z^3)(1+z^2+z^4+z^6)(1+z^4+z^8+z^{12})\cdots$$

$$= \frac{1-z^4}{1-z} \cdot \frac{1-z^8}{1-z^2} \cdot \frac{1-z^{16}}{1-z^4} \cdots = \frac{1}{(1-z)(1-z^2)}$$

and from Example 9.3 this gives $q(r) = \lfloor \frac{r}{2} \rfloor + 1.$

9.21 We have

$$\sum_{r\geq 0} q(r)z^r = \frac{1}{(1-z)(1-z^2)(1-z^4)(1-z^5)} \cdots$$

$$= \frac{1-z^3}{(1-z)(1-z^2)(1-z^3)(1-z^4)(1-z^5)} \cdots$$

$$= (1-z^3) \sum_{r\geq 0} p(r)z^r$$

and the recurrence follows.

9.22 In matrix form, the recurrence is

$$\begin{pmatrix} 1p(1) \\ 2p(2) \\ 3p(3) \\ \cdot \end{pmatrix} = \begin{pmatrix} p(0) \\ p(1) & p(0) \\ p(2) & p(1) & p(0) \\ \cdot & \cdot & \cdot \end{pmatrix} \cdot \begin{pmatrix} \sigma_1(1) \\ \sigma_1(2) \\ \sigma_1(3) \\ \cdot \end{pmatrix} \qquad \begin{pmatrix} \sigma_1(1) \\ \sigma_1(2) \\ \sigma_1(3) \\ \cdot \end{pmatrix} = \begin{pmatrix} 1 \\ 1 & 1 \\ 2 & 1 & 1 \\ \cdot & \cdot & \cdot \end{pmatrix} \begin{pmatrix} \sigma_1(1) \\ \sigma_1(2) \\ \sigma_1(3) \\ \cdot \end{pmatrix} .$$

Finding the initial terms of the inverse matrix (which exists, since $p(0) = 1 \neq 0$) gives the required result.

9.23 The number of paths to (r,r) is $\binom{2r}{r}$ and those that do not rise above the diagonal is (by Theorem 9.36) $\frac{1}{r+1}\binom{2r}{r}$ and hence the result.

9.24 They are enumerated by the Catalan numbers – there is a bijection between mountain ranges and brackets. Replace a left bracket, (, by an upward stroke and a right bracket,), by a downward stroke.

9.25

$$C_r = \frac{1}{(r+1)!} \prod_{k=1}^{r} (4k-2) = \frac{2.6.10.\cdots(4r-2)}{(r+1)!} = \frac{2^r 1.3.5.\cdots(2r-1)}{(r+1)!}$$

$$= \frac{1.2.3.4.5.\cdots(2r-1).(2r)}{(r+1)!r!} = \frac{1}{r+1}\binom{2r}{r}.$$

9.26 The Central Binomial coefficients $\binom{2r}{r}$ have the generating function $\frac{1}{\sqrt{(1-4z)}}$ (see Exercise 9.16). Squaring this, and using the *ogf* product formula, leads to the required result.

Bibliography

[1] Ian Anderson. *A First Course in Discrete Mathematics*. Springer Undergraduate Mathematics Series. Springer-Verlag London Ltd, London, 2001.

[2] Tom M. Apostol. *Introduction to Analytic Number Theory*. Springer-Verlag, New York, 1976. Undergraduate Texts in Mathematics.

[3] Norman L. Biggs. *Codes: an Introduction to Information Communication and Cryptography*. Springer Undergraduate Mathematics Series. Springer-Verlag London Ltd, London, 2008.

[4] Peter J. Cameron. *Combinatorics: Topics, Techniques, Algorithms*. Cambridge University Press, Cambridge, 1994.

[5] Timothy Gowers, June Barrow-Green and Imre Leader, editors. *The Princeton Companion to Mathematics*. Princeton University Press, Princeton, NJ, 2008.

[6] Ronald L. Graham, Donald E. Knuth and Oren Patashnik. *Concrete Mathematics*. Addison-Wesley Publishing Company, Reading, MA, second edition, 1994. A foundation for computer science.

[7] John Haigh. *Probability Models*. Springer Undergraduate Mathematics Series. Springer-Verlag London Ltd, London, 2002.

[8] John M. Howie. *Real Analysis*. Springer Undergraduate Mathematics Series. Springer-Verlag London Ltd, London, 2001.

[9] John M. Howie. *Complex Analysis*. Springer Undergraduate Mathematics Series. Springer-Verlag London Ltd, London, 2003.

[10] Gareth A. Jones and J. Mary Jones. *Information and Coding Theory*. Springer Undergraduate Mathematics Series. Springer-Verlag London Ltd, London, 2000.

A. Camina, B. Lewis, *An Introduction to Enumeration*,
Springer Undergraduate Mathematics Series,
DOI 10.1007/978-0-85729-600-9, © Springer-Verlag London Limited 2011

[11] Imre Lakatos. *Proofs and Refutations*. Cambridge University Press, Cambridge, 1976. The logic of mathematical discovery, Edited by John Worrall and Elie Zahar.

[12] Ivan Niven. Formal power series. *Amer. Math. Monthly*, 76:871–889, 1969.

[13] N. J. A. Sloane. *The On-Line Encyclopedia of Integer Sequences*, http://oeis.org/.

[14] Geoff Smith and Olga Tabachnikova. *Topics in Group Theory*. Springer Undergraduate Mathematics Series. Springer-Verlag London Ltd, London, 2000.

[15] J. H. van Lint and R. M. Wilson. *A Course in Combinatorics*. Cambridge University Press, Cambridge, second edition, 2001.

[16] James N. Webb. *Game Theory*. Springer Undergraduate Mathematics Series. Springer-Verlag London Ltd, London, 2007. Decisions, interaction and evolution.

[17] Herbert S. Wilf. generatingfunctionology. http://www.math.upenn.edu/~wilf.

[18] Herbert S. Wilf. *generatingfunctionology*. A K Peters Ltd., Wellesley, MA, third edition, 2006.

[19] Robin J. Wilson. *Introduction to Graph Theory*. Longman, Harlow, fourth edition, 1996.

A. Camina, B. Lewis, *An Introduction to Enumeration*,
Springer Undergraduate Mathematics Series,
DOI 10.1007/978-0-85729-600-9, © Springer-Verlag London Limited 2011